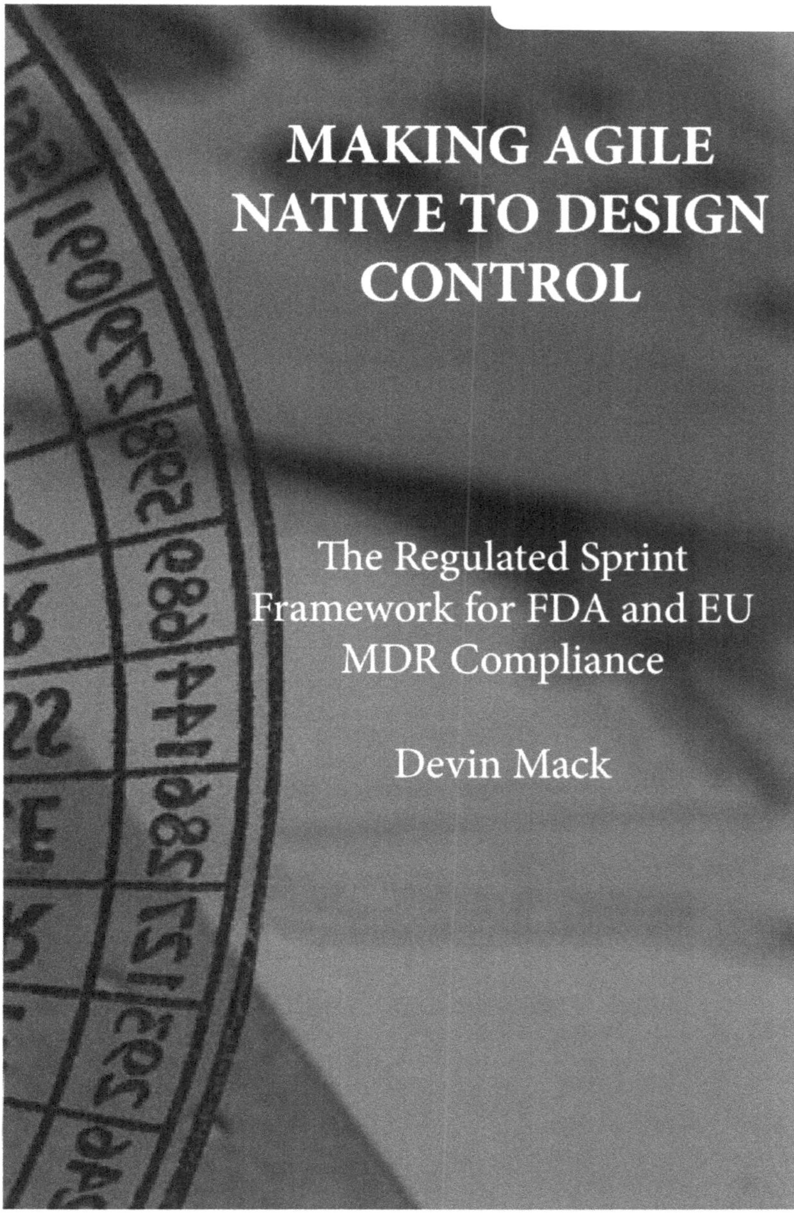

MAKING AGILE NATIVE TO DESIGN CONTROL

The Regulated Sprint Framework for FDA and EU MDR Compliance

Devin Mack

MAKING AGILE NATIVE TO DESIGN CONTROL

The Regulated Sprint Framework for FDA and EU MDR Compliance

Design Quality Engineering

FDA QMSR/21 CFR 820 | ISO 13485:2016 | ISO 14971:2019
IEC 62304:2006+A1:2015 | IEC 62366-1:2015/AMD1:2020
| AAMI TIR45:2023 | AAMI HE75:2025
|EU MDR/IVDR 2017/745 & 2017/746

First Edition: February 2026
Version 1.0
ISBN: 979-8-9947877-0-0
Published by ENQUAL LLC
www.enqual.us

ABOUT THE AUTHOR

Devin Mack

Devin Mack has spent 25+ years as a Design Quality Engineer in regulated product development—medical devices, IVD platforms, drug–device combination products, and the manufacturing interfaces where design intent meets production reality. His hands-on experience spans combination drug delivery systems, surgical instruments, orthopaedic implants, clinical chemistry and immunohematology analyzers, extracorporeal photopheresis platforms, and electro-mechanical-fluidic architectures. The common thread: complex products where design controls, risk management, and verification evidence aren't optional.

He has led and supported Design Controls and DHF execution, ISO 14971 risk management, verification and validation planning, CAPA investigations, change control governance, and design transfer across development and manufacturing. He has built remediation programs in response to FDA warning letters, prepared teams for FDA and notified body inspections, and seen firsthand what happens when governance looks good on paper but breaks under regulatory scrutiny. That experience—watching systems fail under inspection pressure—is the foundation of this book.

An ASQ Six Sigma Black Belt, Devin brings statistical rigor to design verification and quality decision-making. His approach is simple: compliance should be produced by the development system during execution, not reconstructed after the fact. If your process requires a separate "compliance sprint" to assemble evidence, your process is broken.

Through ENQUAL LLC, Devin works with organizations modernizing their development practices under FDA and EU regulatory frameworks—integrating agile and iterative execution into design-control environments so traceability, risk evidence, and verification outcomes are generated as natural byproducts of the work.

Making Agile Native to Design Control is the operational blueprint he wished existed when he started solving this problem.

Preface: Who This Book Is For

This book is for anyone who has felt the tension between moving fast and staying compliant.

If you're an engineering leader who watches your team finish a sprint, then spend two weeks 'catching up' on documentation, this book is for you.

If you're a QA or regulatory professional who feels like you're always chasing engineering, asking for documents that should have been created weeks ago—this book is for you.

If you're a program manager trying to explain to leadership why releases keep slipping even though 'development is done'—this book is for you.

If you've ever sat in an FDA inspection and wished you could show exactly how a change flowed through your system, with timestamps that prove you did it right—this book is definitely for you.

If you're navigating dual-market compliance—FDA and EU MDR/IVDR—and need a single framework that satisfies both without parallel quality systems, this book gives you that foundation.

A Note on Regulatory Citations: Throughout this book, design control requirements are cited to 21 CFR 820.30 for precision and historical continuity. Under QMSR (effective February 2, 2026), these requirements are incorporated through ISO 13485:2016 Clause 7.3. The substantive requirements are unchanged; only the citation pathway differs. See Appendix TT for complete QMSR mapping.

What This Book Is Not

This is not a book about Agile methodology. There are plenty of those.

This is not a book that explains ISO 14971 or IEC 62304 clause by clause. The standards themselves do that.

This is not a theoretical framework that sounds good in a presentation but falls apart when you try to implement it.

What This Book Is

This is a practical framework for making design control feel like part of your engineering workflow rather than something bolted on afterward.

Every concept in this book has been tested in real organizations, under real regulatory scrutiny, with real inspectors asking real questions.

The framework you'll learn isn't new work—it's a different way of organizing work you already do. That's the key insight that changes everything.

A Critical Clarification

This framework does not remove hazards, replace risk management, or eliminate regulatory accountability. It changes when evidence is created—not what evidence is required. Every verification activity, every risk control, every traceability requirement mandated by 21 CFR 820.30, ISO 14971, and IEC 62304 remain fully in force. What changes is the timing: evidence emerges during development rather than being retrofitted afterward. The regulatory burden is identical. The compliance outcome is identical. Only the workflow integration differs.

QMSR and CP 7382.850: Why This Book Matters Now

On February 2, 2026, FDA's Quality Management System Regulation (QMSR) took effect, replacing 21 CFR Part 820 with ISO 13485:2016 incorporated by reference. The same day, FDA's updated Compliance Program CP 7382.850 provides a primary inspection framework for QMSR-era medical device inspections. These changes are not cosmetic. They represent FDA's clear expectation for integrated, risk-based quality systems—exactly what this framework implements.

CP 7382.850 organizes inspections around six QMS Areas—Change Control, Design and Development, Management Oversight, Measurement Analysis and Improvement, Outsourcing and Purchasing, and Production and Service Provision—with "risk" at the center. Investigators follow the flow of changes through your system, asking: "Show me how a design change moved through your system." That single question is the entire heart of this book. Your regulated sprint loop, your CCB classification, your parallel execution model, your Sprint DHF packets—they answer that question with timestamps, traceability, and evidence that emerged during development, not after.

This framework was not retrofitted for QMSR. It was built from first principles that happen to match what FDA is trying to accomplish with harmonization. The alignment is structural, not cosmetic. See Appendix TT for detailed QMSR mapping and Appendix UU for clause-by-clause ISO 13485 alignment. The framework also addresses EU MDR 2017/745 and EU IVDR 2017/746 requirements, enabling organizations to achieve dual-market compliance through a single regulated sprint process. See Appendix LLL for comprehensive EU alignment.

How This Book Is Organized

This book follows a deliberate progression:

Part I shows you the problem—why SaMD teams feel like Agile and design control are fighting each other, and what that costs.

Part II reveals the insight—the single realization that changes everything about how you think about compliance.

Part III establishes the foundation—precise terminology and role definitions that make the framework work.

Part IV presents the framework itself—the regulated sprint, the risk gate, parallel execution, and the tracking system.

Part V addresses practical implementation—how to integrate usability engineering and use-related risk, manage post-market surveillance across FDA and EU MDR requirements, define release gates, and handle emerging signal types including AI/ML and cybersecurity.

Part VI prepares you for inspection—how to demonstrate confidence when FDA investigators or EU Notified Body auditors ask questions, and how to avoid the common failure modes that undermine frameworks.

Part VII looks at the future—how this framework evolves with the industry.

Read the parts in order the first time. After that, use the appendices as reference.

ABBREVIATIONS AND ACRONYMS

AAMI — Association for the Advancement of Medical Instrumentation

AI — Artificial Intelligence

AI/ML — Artificial Intelligence / Machine Learning

API — Application Programming Interface

CAPA — Corrective and Preventive Action

CCB — Change Control Board

CDS — Clinical Decision Support

CFR — Code of Federal Regulations

CI/CD — Continuous Integration/Continuous Deployment

CP — Compliance Program (FDA)

CSA — Computer Software Assurance

CVE — Common Vulnerabilities and Exposures

DHF — Design History File

DMR — Device Master Record

DQE — Design Quality Engineer

EU MDR — European Union Medical Device Regulation

EUDAMED — European Database on Medical Devices

FDA — Food and Drug Administration

FMEA — Failure Mode and Effects Analysis

FTA — Fault Tree Analysis

GMLP — Good Machine Learning Practice

GSPR — General Safety and Performance Requirements (EU MDR/IVDR)

HA — Hazard Analysis

HE75 — AAMI Human Factors Engineering standard

IEC — International Electrotechnical Commission

IFU — Instructions for Use

ISO — International Organization for Standardization

IVDR — In Vitro Diagnostic Regulation (EU Regulation 2017/746)

MDR — Medical Device Reporting (FDA, 21 CFR Part 803; distinct from EU MDR)

ML — Machine Learning

NB — Notified Body

NC — Nonconformance

OTS — Off-The-Shelf (software)

P0 — Probability of error causing the hazardous situation (used in P1 decomposition)

P1 — Probability of hazardous situation occurring

P2 — Probability of harm given hazardous situation

PCCP — Predetermined Change Control Plan

PM — Program Manager / Product Manager

PMCF — Post-Market Clinical Follow-up

PMS — Post-Market Surveillance

PO — Product Owner

PRRC — Person Responsible for Regulatory Compliance (EU MDR Article 15)

PSUR — Periodic Safety Update Report

QA — Quality Assurance

QMS — Quality Management System

QMSR — Quality Management System Regulation

RA — Regulatory Affairs

RACI — Responsible, Accountable, Consulted, Informed

RPN — Risk Priority Number

SaMD — Software as a Medical Device

SBOM — Software Bill of Materials

SDP — Software Development Plan

SDS — Software Design Specification

SME — Subject Matter Expert

SOP — Standard Operating Procedure

SOUP — Software of Unknown Provenance

SRS — Software Requirements Specification

TC — Test Case

TIR — Technical Information Report

UDI — Unique Device Identification

UDI-DI — UDI Device Identifier (static portion of the UDI)

UDI-PI — UDI Production Identifier (variable portion of the UDI)

UEF — Usability Engineering File

UI — User Interface

UOUP — User Interface of Unknown Provenance (legacy UI without complete development records)

URRA — Use-Related Risk Analysis

UX — User Experience

V&V — Verification and Validation

Table of Contents

How to Read This Book

This book is designed for three types of readers. Choose the path that matches your immediate need, then explore the rest at your own pace.

Path 1: Executive Overview (2 hours)

For engineering leaders, VPs, and program managers who need to understand the framework and make a go/no-go decision. Read Preface → Part I (Chapters 1–3) → Chapter 4 → Chapter 8 → Chapter 19 → Epilogue. Then skim Appendix K (Engineering Leader Quick Reference).

Path 2: Implementer (1–2 weeks)

For DQEs, scrum masters, QA/RA professionals, and development leads who will configure and run the framework. Read Parts I through V sequentially. Then work through Appendix A (90-Day Implementation), Appendix B (Bug Walkthrough), Appendix H (HIGH Risk Feature Walkthrough), Appendix M (Sprint DHF Packet Template), and Appendix FF (Jira Configuration). Use the Design Control Tracker in Appendix GGG as your working reference.

Path 3: Inspection Defense (pre-audit)

For teams preparing for an FDA inspection or responding to a 483 observation. Start with Part VI (Chapters 17–18). Then read Appendix F (483 Pattern Library), Appendix G (Mock Inspection Q&A), Appendix GG (Extended Mock Inspection Scenarios), Appendix OO (Inspection Preparation Deep Dive), and Appendix X (Audit Preparation Checklist). Return to Parts III–IV to understand the framework that supports your defense.

Key Terms Used in This Book

This book is precise about language because language drift causes audit failures (Chapter 6 explains why). To prevent misreading before you reach those definitions, here are the terms that matter most:

Safety_Impact (HIGH / MEDIUM / LOW): A governance classification set by the CCB that determines the rigor of design controls applied to a change. It is aligned to risk-based thinking but is not the full ISO 14971 risk estimate (combination of severity and probability of occurrence of harm). Think of it as: how much evidence does this change require?

Safety_Impact routes rigor; ISO 14971 evaluates risk in the risk file using Hazard → Hazardous Situation → Harm with P1/P2.

Hazard → Hazardous Situation → Harm: The ISO 14971 causal chain. A hazard is a characteristic of the device (not a user action). A hazardous situation occurs when a person is exposed to that hazard. Harm is the resulting injury or damage. This sequence matters because auditors follow it.

CCB (Change Control Board): The governance gate where every change is classified before entering a sprint. No exceptions. The CCB assigns Safety_Impact and determines what controls apply.

DQE (Design Quality Engineer): The role responsible for ensuring design controls are followed. Not QA (who inspects product) and not RA (who writes submissions). The DQE owns the process.

Retrofit Tax: The cost of recreating evidence after the fact. When teams develop first and document later, they pay this tax—in time, accuracy, and audit risk. This book eliminates it by generating evidence during development.

Full definitions with regulatory context appear in Chapter 6 (The Language That Changes Everything). If you encounter an unfamiliar term before then, check Appendix E (Glossary).

The Destination Before the Journey

Before you begin, here is what the complete framework looks like. This is the Regulated Sprint process—the integrated loop where development, risk management, and design control evidence move together. Parts I through III explain why this structure works. Part IV shows you how to build it. Study this figure now so you recognize the destination as the book explains the journey.

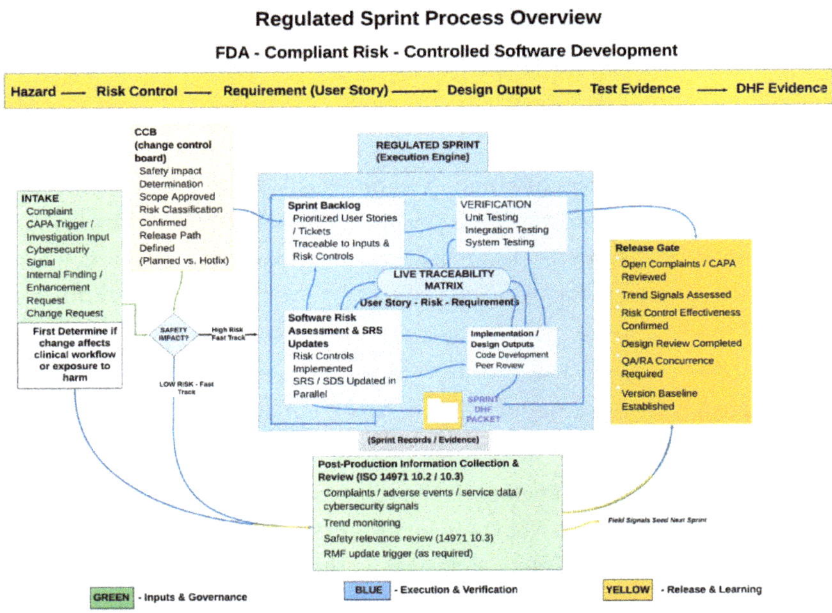

Figure 1: The Regulated Sprint Process Overview (detailed walkthrough in Chapter 8)

PART I — THE PROBLEM

Before we can fix something, we need to understand exactly why it's broken.

Chapter 1 — The Pain Every SaMD Team Knows

You've felt it.

The sprint ends. Engineering says 'done.' And then someone from Quality asks: 'Where's the updated risk file? Where's the traceability? Where's the verification evidence that ties back to the requirement?'

The team groans. They just finished building something. Now they have to go back and document what they built. Again.

This is the experience of nearly every software team developing medical devices. It feels like there are two jobs: building the software, and then proving you built it correctly. The first job is engineering. The second job feels like bureaucracy.

1.1 The Symptoms

You know you have this problem if any of these sound familiar:

Sprint velocity looks great, but releases keep slipping because 'documentation isn't ready.' Teams celebrate completing features, then discover weeks of compliance work remains before those features can ship.

QA always seems to be 'catching up' to where engineering was two weeks ago. There's a permanent lag between what's built and what's documented.

Risk files get updated in batches, usually right before audits or submissions. The timestamps cluster suspiciously around external deadlines rather than development activity.

Engineers view design control as something QA does, not something they do. When asked about requirements or risk, they point to another team.

The phrase 'engineering-only change' gets used to justify skipping process. Refactoring, performance optimization, and infrastructure work somehow exist outside the quality system.

Release gates feel like the first time anyone really checks if things are ready. Surprises are common. Rework is expected.

> **483 PATTERN:** Documentation timestamps cluster around submission dates rather than development dates. Inspectors see risk analysis dated weeks after code commits, revealing reconstruction rather than concurrent development.

1.2 The Retrofit Tax

Organizations rarely measure the retrofit tax because it's spread across so many activities. But consider:

How many hours does your team spend after each sprint updating documents to match what was built? This includes SRS updates, risk file updates, traceability matrix updates, and verification summary updates.

How many release delays come from discovering gaps in traceability, risk analysis, or verification? Each gap discovered at the release gate represents work that should have happened during the sprint.

How much time do you spend preparing for audits by reconstructing evidence that should have existed all along? The 'audit prep' scramble is a symptom of evidence not being captured contemporaneously.

For most organizations, the retrofit tax adds 20-40% to the total effort of each release. Not because design control requires that much work—but because doing it afterward is far more expensive than doing it during.

> **KEY INSIGHT:** The retrofit tax isn't the cost of compliance. It's the cost of compliance done at the wrong time.

1.3 The Weekly Cycle of Pain

Here's how the retrofit cycle typically plays out:

Week 1-2 of a sprint: Engineering builds features. Product refines requirements. Testing happens. Everything feels productive. The standup is optimistic.

Sprint ends: Engineering moves to the next sprint. The 'done' features enter a twilight zone where they're technically complete but not releasable. Engineers have mentally moved on.

Week 3-4: QA scrambles to create or update the risk file, trace requirements to tests, document design decisions that were made verbally in standups. Engineers get pulled back to answer questions about code they've already mentally moved on from. Context is lost. Frustration builds.

Release prep: Someone realizes the traceability matrix has gaps. Test evidence doesn't clearly tie to requirements. Risk controls reference requirements that changed during implementation. More rework. More frustration.

This cycle repeats every sprint. Teams get so used to it that they build buffer time into their schedules. They've normalized the dysfunction.

> **483 PATTERN:** Risk file shows bulk update on single date after code freeze. Inspector asks why risk management wasn't concurrent with development. Team cannot explain because it wasn't—risk analysis was reconstruction, not real-time assessment.

1.4 The Organizational Friction

The retrofit tax creates organizational friction that goes beyond schedule delays:

Engineering and QA develop adversarial dynamics. Engineers see QA as a blocker. QA sees engineers as people who don't follow process. Neither is wrong from their perspective—the system pits them against each other.

Talented engineers leave for companies where they can 'just build.' They didn't sign up for medical devices to spend 40% of their time on documentation. They wanted to build software that helps patients.

Quality professionals burn out. They're always chasing, always catching up, always the bearers of bad news. The job becomes about policing rather than enabling.

Leadership loses trust in schedules. When 'development complete' doesn't mean 'ready to ship,' planning becomes guesswork. Commitments to customers become unreliable.

> **ROLE ACCOUNTABILITY:** In retrofit organizations, no single role owns the gap between 'code complete' and 'release ready.' This ownership vacuum is the root cause of the friction.

Chapter 1 Summary

The pain is real: sprint velocity that doesn't translate to releases, documentation that lags development, organizational friction between engineering and quality.

The cost is measurable: 20-40% overhead on every release, plus unmeasured costs in turnover, trust, and morale.

The pattern is recognizable: work happens, then documentation happens, then gaps are discovered, then rework happens.

The question is: does it have to be this way?

The retrofit tax is ultimately a timestamp problem: evidence created after the fact describes what was done, but cannot prove what was known when the decision was made. Auditors are trained to read that gap.

CAUSAL CHAIN PRIMER: WHY THE LANGUAGE MATTERS

Every risk in this book follows the same three-level structure from ISO 14971:

Hazard: the device characteristic that creates potential for harm.

Example: 'Software UI allows selection of infusion rate without hard-stop validation'

Hazardous Situation: the exposure event, including use error / incorrect interaction.

Example: 'Clinician selects 10× intended dose due to ambiguous unit display'

Harm: the clinical outcome.

Example: 'Overdose causing respiratory depression requiring intervention'

Use error belongs at the Hazardous Situation level — never at the Hazard level. This distinction determines how you write risk files, how you classify changes, and how you answer an inspector who asks: 'What is the hazard for this risk?'

Chapter 2 — Why Agile and Compliance Feel Like Enemies

The frustration is real, and it has a source.

Agile development emphasizes iteration, adaptation, and working software over comprehensive documentation. Design control emphasizes traceability, documented evidence, and proof that requirements were met.

On the surface, these seem incompatible. One says 'move fast and adapt.' The other says 'document everything and prove it.'

But here's what most organizations miss: the conflict isn't between the methodologies. The conflict is in how organizations have chosen to implement them.

2.1 The False Choice

Many organizations have structured themselves around a false choice:

Option A: Move fast with Agile, then bolt on compliance afterward. This is the retrofit model. It preserves development agility but creates the documentation lag and all its associated costs.

Option B: Do compliance 'properly' by slowing everything down. This is the waterfall-in-disguise model. Requirements must be locked before development. Design must be frozen before implementation. Every change requires a full impact assessment before work can begin.

Neither option works. Option A creates the retrofit tax. Option B destroys the benefits of iterative development and still doesn't guarantee compliance—it just moves the dysfunction earlier in the process.

The industry has been arguing about which option is 'right' for over a decade. The argument is pointless because both options are wrong.

2.2 What Regulators Actually Want

Regulators don't care about your sprint methodology. They care about evidence.

They want to see that when you made a design decision, you understood its safety implications.

They want to see that when you changed a requirement, you updated the risk analysis.

They want to see that when you verified functionality, you knew which hazards that verification was controlling.

Notice what's missing from that list: regulators don't specify when this work should happen. They don't say 'do risk analysis before coding' or 'do risk analysis after coding.' They say the risk analysis must exist and must be current.

> **REGULATORY ANCHOR:** 21 CFR 820.30(g) requires that 'design validation shall ensure that devices conform to defined user needs and intended uses.' It does not prescribe the timing of this validation relative to development activities. (Under QMSR, effective February 2, 2026, this requirement maps to ISO 13485:2016 Clause 7.3.6; the substantive obligation is unchanged.)

> **REGULATORY ANCHOR:** ISO 14971:2019 Clause 4.1 requires that risk management be 'a continual process throughout the life cycle.' This explicitly describes ongoing activity, not a phase gate.

The timing problem is entirely self-inflicted. Organizations chose to separate development from compliance. Regulators never required that separation.

2.3 The Organizational Divide

In most organizations, there's a clear divide:

Engineering teams work in sprints, use Jira, track velocity, and consider stories 'done' when code is merged and tests pass. Their world is features, bugs, and technical debt.

Quality teams work in document management systems, track compliance status, and consider changes 'done' when documentation is approved. Their world is DHF packets, risk files, and traceability matrices.

These two systems rarely talk to each other in real time. Engineering's 'done' happens in week 2. Quality's 'done' happens in week 6. The gap between them is the retrofit tax.

The gap exists because organizations built two separate systems for what should be one unified activity. Development and compliance aren't two jobs—they're two perspectives on the same job.

2.4 The Permission Myth

Many organizations believe they need 'permission' to do Agile in a regulated environment. They wait for FDA to explicitly say 'Agile is okay.' They look for guidance that tells them exactly how to structure sprints for compliance.

This waiting is unnecessary. The regulations have always allowed iterative development. Nothing in 21 CFR 820 requires waterfall. Nothing in ISO 14971 requires sequential phases. Nothing in IEC 62304 prohibits sprints. The regulations and standards define required outcomes and evidence—they do not mandate a specific development lifecycle sequence.

AAMI TIR45:2023—a consensus standard acknowledged by FDA—makes this explicit. But TIR45 didn't create new permissions. It documented what was already allowed.

> **REGULATORY ANCHOR:** AAMI TIR45:2023 establishes that iterative and Agile development methodologies are compatible with IEC 62304 requirements. FDA's acknowledgment of TIR45 provides regulatory 'air cover' for organizations implementing these approaches.

> **KEY INSIGHT:** Agile and design control aren't enemies. They're the same system viewed from different angles. The conflict only exists when you separate them.

INSPECTION SIDEBAR: When Asked About Agile Compatibility

Weak answer: "We follow Agile but add design control activities at milestones."

Strong answer: "We've integrated design control into our sprint cadence. Every change is classified before sprint entry, evidence is captured during the sprint, and the sprint DHF packet is complete when the sprint ends. Our approach follows AAMI TIR45 guidance, which FDA has acknowledged."

Chapter 2 Summary

The conflict between Agile and design control is organizational, not inherent. Regulators don't require sequential phases, locked requirements, or separate compliance systems. Organizations created those constraints themselves.

The regulations describe what must be done, not when. 21 CFR 820.30, ISO 14971, and IEC 62304 all allow iterative development. AAMI TIR45:2023 makes this explicit.

The false choice between 'move fast' and 'be compliant' is a false choice. Both options fail. The answer is neither—it's integration.

Chapter 3 — The Inspection That Reveals Everything

Nothing exposes the retrofit problem like an FDA inspection or a Notified Body audit.

Inspectors are trained to look for one thing above all else: evidence that your system actually works the way you say it does. And they have a devastatingly simple way to test this.

3.1 The Timestamp Test

An experienced inspector will pick a change—any change—and trace it through your system. They'll look at:

When was the requirement documented? When was the risk analysis updated? When was the code committed? When was the verification performed? When were these activities approved?

If your system works the way it should, these timestamps will be interleaved. Requirements before or during implementation. Risk analysis updating as design solidifies. Verification after implementation but before the sprint ends. All within the same general timeframe.

If your timestamps tell a different story—all the documentation dated weeks after the code, or all approvals clustered on the same day—the inspector knows. They've seen it hundreds of times.

> **483 PATTERN:** Risk file updates dated 3-4 weeks after corresponding code commits. This pattern indicates reconstruction rather than concurrent risk management. Inspectors will ask: 'If you didn't update the risk file until three weeks later, how did you know the design was safe when you made the decision?'

3.2 The Question You Can't Answer

Here's the question that trips up organizations with retrofit compliance:

> *Show me the sprint where this change was implemented, and show me the risk analysis that was current during that sprint.*

If your risk analysis was updated after the sprint, you can't answer this question. You can show the risk analysis that exists now. But you can't show what you knew when you made the decision.

This matters because design control isn't just about having documents. It's about having documents that demonstrate understanding at the point of decision. The document that proves you understood the risk after shipping doesn't prove you understood it before.

> **483 PATTERN:** When asked 'what was the risk classification when this was implemented?', team cannot answer because classification happened retroactively. The CCB record—if it exists—postdates the implementation.

3.3 The Two Organizations

Consider two organizations responding to the same inspector question:

Inspector: 'Walk me through how this algorithm change was controlled.'

Organization A (Retrofit Compliance)

'We implemented the change in Sprint 47. Then our QA team updated the risk file and traceability matrix. Here's the current documentation.'

Inspector follow-up: 'When was the risk file updated?'

'About three weeks after the sprint ended.'

Inspector follow-up: 'So during Sprint 47, what risk analysis guided the implementation?'

[Hesitation] 'We... used the previous version of the risk file.'

Inspector follow-up: 'But that version didn't include this algorithm change. How did engineers know which hazards they were addressing?'

[Silence. No answer available.]

Organization B (Native Compliance)

'Every change goes through our CCB before entering a sprint. This change was classified as medium risk on March 3rd—here's the CCB record. During Sprint 47, our DQE updated the risk file on March 8th as the design solidified—here's the revision history. Verification was completed March 12th with DQE approval—here's the test evidence with timestamp. Here's the Sprint 47 DHF packet with the complete evidence chain.'

INSPECTION Q&A: P1/P2 SEPARATION

Inspector: 'How did you estimate the probability that this hazardous situation occurs?' (This is asking for P1 — exposure probability.)

Strong answer: 'We used usability study data from formative evaluations to estimate how frequently users are exposed to this interaction pathway. That exposure probability feeds P1 in our risk estimate. IEC 62366-1 URRA refined our P1; it did not replace our ISO 14971 risk analysis.'

Inspector: 'Given that exposure — given the hazardous situation occurs — why is the probability of harm acceptable?' (This is asking for P2 — harm probability given exposure.)

Strong answer: 'P2 is governed by our ISO 14971 risk analysis. Here is the clinical rationale for the harm probability we assigned, and here is the risk control that reduced residual P2 to within our acceptance criteria.'

Organizations that conflate P1 and P2 cannot answer these questions separately — and the inability to answer them separately is an inspection finding.

Inspector follow-up: 'Show me the risk file update from March 8th.'

'Here it is. You can see the hazard analysis was updated to reflect the new algorithm boundary conditions. The engineer who made the design decision and the DQE who updated the risk file were working in the same sprint, talking daily.'

The difference isn't that Organization B has better documentation. The difference is that Organization B did the work at the right time, and their timestamps prove it.

3.4 What Inspectors Are Really Looking For

Inspectors aren't trying to catch you in a 'gotcha.' They're trying to assess whether your system produces safe devices reliably.

A system that produces concurrent evidence—risk analysis updated during design, verification linked to current requirements, approvals at the point of completion—gives inspectors confidence. Even if there are minor gaps, the system is working.

A system that produces reconstructed evidence—documentation created after the fact, timestamps that don't align with development—raises questions. Even if the documents are complete, the process is suspect.

The difference isn't the quality of your documentation. It's the story your timestamps tell.

> **KEY INSIGHT:** Evidence created during development demonstrates understanding. Evidence created afterward demonstrates compliance effort. Inspectors can tell the difference.

INSPECTION SIDEBAR: When Asked to Show Your System

> **Weak answer:** "Here are our procedures and our document repository."
>
> **Strong answer:** "Let me walk you through a specific change end-to-end. Here's when it entered our system, here's who classified it, here's the sprint where we implemented it, here's the evidence generated during that sprint. The timestamps tell the story."

Chapter 3 Summary

Inspections reveal the truth that internal processes can hide. The timestamp test exposes whether evidence was created during development or reconstructed afterward.

Organizations with retrofit compliance cannot answer the fundamental question: 'What did you know when you made the decision?'

Organizations with native compliance can trace any change through the complete cycle with timestamps that prove concurrent execution.

The goal isn't to survive inspections—it's to build a system where inspections are easy because the evidence naturally exists.

PART II — THE INSIGHT

The breakthrough isn't doing more. It's doing the same work at a different time.

Chapter 4 — The Temporal Insight

Here is the insight that changes everything:

> *The conflict between Agile and design control is not philosophical. It is temporal.*

The problem isn't that Agile wants one thing and design control wants another. The problem is when design control activities happen relative to development activities.

This single realization—that timing is the issue, not methodology—unlocks everything else in this book.

4.1 Before, During, After

There are only three times when design control can happen:

BEFORE Development
Requirements are locked. Risk analysis is complete. Test plans are approved. Then—and only then—development begins.

This is the traditional 'waterfall' approach. It provides excellent documentation but destroys the ability to learn and adapt during development. By the time you discover that a requirement doesn't work, you've already committed to it. Change becomes expensive because every change requires revisiting the entire upstream documentation chain.

Organizations that attempt this approach in iterative environments end up with one of two outcomes: either they abandon the approach under schedule pressure, or they create elaborate fictions where documents are 'approved' but everyone knows they'll change.

AFTER Development
Code is written. Features are built. Then the compliance team goes back and documents what happened, updates the risk file, creates traceability.

Temporal integrity is what makes your benefit-risk conclusion defensible. A benefit-risk determination supported by contemporaneous evidence — risk analysis updated during design, verification linked in real time, release decision grounded in evidence that existed at the moment of decision — is what regulators describe as objective. Later reconstruction may produce the same words, but it cannot produce the same timestamps, and timestamps are what inspectors read first.

This is the retrofit approach. It preserves development agility but creates the timestamp problem. Your documentation proves you're compliant now, but not that you were thinking about compliance when you made decisions.

More insidiously, retrofit documentation often doesn't accurately reflect what actually happened. The QA professional updating the risk file three weeks later is reconstructing decisions from memory, code comments, and Jira tickets. Details are lost. Rationale is invented. The documentation describes what should have happened, not what did happen.

> **483 PATTERN:** Risk analysis contains generic hazards that don't match specific design decisions. Inspector recognizes that risk analysis was written by someone who didn't participate in design—it reads like a template, not a record of actual engineering judgment.

DURING Development
Requirements evolve as understanding grows. Risk analysis updates as design solidifies. Testing happens alongside implementation. Everything moves together.

This is native design control. It's not a compromise between the other two approaches—it's a recognition that they both solve the wrong problem.

When design control happens during development, the documentation is accurate because it's created by the people doing the work, at the time they're doing it. The timestamps are interleaved because the activities are interleaved. The evidence demonstrates understanding because it was created at the moment of understanding.

4.2 Why 'During' Changes Everything

When design control happens during development, several things change:

Engineers don't experience compliance as extra work. Risk analysis isn't something that happens after they're done—it's part of how they think about design decisions. When the DQE asks 'what hazardous situations does this create?', that question shapes the design, not just the documentation.

QA doesn't have to chase engineering. They're in the room when decisions are made, updating documents in real time. There's no reconstruction because there's nothing to reconstruct—the evidence was captured as it happened.

Timestamps tell the right story. Evidence is created at the point of understanding, not manufactured later. An inspector can see that risk analysis was updated on the same day as a design decision, not three weeks afterward.

Releases don't have a 'documentation catch-up' phase. The documentation is already complete because it was created as part of the work. When the sprint ends, the design history is current.

4.3 The Same Work, Different Timing

Here's what's crucial to understand: native design control doesn't require more work than retrofit compliance. In fact, it typically requires less.

The work is the same:

Table 4-1: Retrofit vs. Native Timing — Same work, different timing

Activity	Retrofit Timing	Native Timing	Work Difference
Requirements documentation	After design solidifies	As design evolves	Same work, less rework
Risk analysis	After implementation	During design decisions	Same work, more accuracy

Activity	Retrofit Timing	Native Timing	Work Difference
Design documentation	After code complete	During architecture	Same work, captured live
Verification	After sprint	During sprint	Same work, faster feedback
Traceability	Before release	Continuous	Same work, no scramble

The difference is who does it and when.

In retrofit compliance, QA does most of this work after development, reconstructing what engineers did. This is slow because QA has to investigate what happened and translate it into compliance artifacts. Context is lost. Engineers are pulled back to answer questions. Friction accumulates.

In native compliance, the sprint team does this work during development, capturing it as it happens. This is faster because the information is fresh and the people with context are doing the documentation. No reconstruction. No investigation. No friction.

> **KEY INSIGHT:** Native design control isn't additional work bolted onto Agile. It's recognition that you're already doing design control—you're just not capturing it in a form regulators recognize.

4.4 Evidence as Byproduct

The ultimate goal is evidence as byproduct, not evidence as artifact.

In retrofit compliance, evidence is an artifact—something you create to prove you did something. It's separate from the work itself. Creating evidence feels like overhead because it is overhead.

In native compliance, evidence is a byproduct—a natural output of how work is done. You don't create evidence separately; evidence emerges from the process. Timestamps appear because activities have timestamps. Traceability exists because links are created as part of workflow. Risk analysis is current because it's updated when designs change.

When evidence is a byproduct, compliance stops feeling like compliance. It's just how work gets done.

MOCK INSPECTION Q&A: When Asked About Evidence Creation

> **Weak answer:** "We create our documentation package before each release."
>
> **Strong answer:** "Evidence is generated as work happens. When an engineer makes a design decision, the DQE updates the risk file that day. When tests pass, the results are linked to requirements immediately. By sprint end, the evidence already exists—we're not creating it, we're confirming it."

Chapter 4 Summary

The conflict between Agile and design control is temporal, not philosophical. The question isn't whether to do design control—it's when.

Before development: documentation-heavy, change-resistant, ultimately fictional in iterative environments.

After development: agile-friendly but evidence is reconstructed, timestamps reveal the truth, inspectors notice.

During development: work and evidence happen together, timestamps interleave naturally, compliance is a byproduct.

The rest of this book shows you how to make 'during' practical.

Chapter 5 — Design Control Was Never the Enemy

If you've been fighting design control, here's something that might surprise you: the regulations aren't the problem.

Read 21 CFR 820.30 carefully. Read ISO 14971. Read IEC 62304. None of them say:

'Complete all requirements before starting development.'

'Freeze the design before implementation.'

'QA must approve changes before engineering can work on them.'

'Separate your compliance system from your development system.'

These constraints exist in many organizations, but they weren't imposed by regulators. Organizations created them, often in misguided attempts to 'ensure compliance.'

5.1 What the Regulations Actually Say

> **REGULATORY ANCHOR:** 21 CFR 820.30(c) requires 'design input requirements relating to a device.' It does not require these requirements to be complete before development begins. It requires them to be documented and appropriate.

> **REGULATORY ANCHOR:** ISO 14971:2019 requires that risk management be conducted 'throughout the life cycle.' This is explicitly not a phase-gate model—it's a continuous process that evolves with the design.

> **REGULATORY ANCHOR**: IEC 62304:2006+A1:2015 (Edition 1.1) defines software development processes by safety class (A, B, C). For Class B and C software, it requires risk-based development. Amendment 1 (2015) notably extended system testing requirements (Clause 5.7) to all classes including Class A. The standard does not prescribe waterfall or any specific methodology.

> **REGULATORY ANCHOR:** ISO 13485:2016 Clause 7.3.2 on Design and Development Planning states that 'planning output shall be

updated, as appropriate, as the design and development progresses.' This explicitly anticipates iterative refinement.

The regulations describe what must be done, not when. The timing choices that create conflict are organizational decisions, not regulatory requirements.

5.2 Where the Constraints Actually Come From

If regulators didn't create these constraints, who did?

Often, they came from well-intentioned quality professionals who interpreted 'documented requirements' as 'locked requirements' and 'risk management throughout the life cycle' as 'risk management before development starts.'

Design control is the system that ensures risk acceptability decisions are made with objective evidence, not opinion. When an engineer decides a change is low risk, design control is the mechanism that records what evidence supported that decision at the moment it was made. Without that record, the decision cannot be defended — not because it was wrong, but because the evidence of its correctness does not exist.

These interpretations weren't unreasonable—they reflected how medical device development worked when most products were hardware. For a pacemaker or an implant, you really do need to lock specifications before manufacturing begins. You can't iterate on a device that's already been implanted in a patient.

But software doesn't work that way. Software can be updated. Software benefits from iteration. Software development naturally produces the evidence regulators want—if you capture it.

Sometimes constraints came from responses to audit findings. An inspector asked 'where was this requirement documented?' and the organization responded by requiring all requirements to be documented and approved before development could begin. The cure was worse than the disease.

Sometimes constraints came from consulting advice. Well-meaning consultants, trained in traditional medical device development, applied hardware thinking to software projects. 'Best practices' calcified into mandatory procedures.

The result is organizations that are more constrained than regulations require, and blame the regulations for constraints they imposed on themselves.

> **483 PATTERN:** Organization cites 'regulatory requirement' for practices that have no regulatory basis. When asked to point to the specific clause, they cannot. Their constraints are policy choices, not compliance obligations.

5.3 The Permission You Already Have

AAMI TIR45:2023—a consensus standard acknowledged by FDA—establishes that iterative and Agile development methodologies are compatible with IEC 62304 requirements.

This isn't news. It's been true all along. TIR45 simply made it explicit for organizations that needed regulatory 'permission' to do what the regulations already allowed.

TIR45 provides specific guidance on how to maintain compliance in iterative environments:

Requirements can evolve during development as long as they're documented before verification completes.

Risk analysis can be updated iteratively as design understanding grows.

Verification can happen throughout development, not just at the end.

The 'Design History File' can be assembled from sprint artifacts rather than created as a separate deliverable.

> **REGULATORY ANCHOR:** AAMI TIR45:2023 Section 5.2 states: 'The medical device software development life cycle processes in IEC 62304 do not require any specific software development model, nor

do they explicitly preclude the use of any specific development model.'

5.4 What Good Engineering Looks Like

Here's a secret that experienced engineers know: design control, properly implemented, is just good engineering practice with documentation.

Good engineers think about requirements before coding. Design control asks you to write them down.

Good engineers consider what could go wrong. Design control asks you to document your risk thinking.

Good engineers verify that their code works. Design control asks you to keep the test results.

Good engineers know why they made design decisions. Design control asks you to record the rationale.

The documentation isn't overhead—it's the evidence that good engineering happened. If your team already does good engineering, making it visible to regulators shouldn't require fundamental changes to how you work.

If your team doesn't do good engineering—if they code without understanding requirements, ignore risks, skip testing, and make decisions without rationale—then design control will be painful. But the pain isn't caused by design control. Design control is revealing problems that already exist.

> **KEY INSIGHT:** Design control was never the enemy. It was a set of reasonable requirements that organizations implemented in unreasonable ways. This book shows the reasonable way.

5.5 The Mental Shift

The mental shift required is from 'design control as compliance activity' to 'design control as engineering practice.'

Compliance activity: Something we do because regulators require it. Extra work. Overhead. The tax we pay to sell medical devices.

Engineering practice: Something we do because it produces better outcomes. Core work. Value-added. How we build safe, effective software.

When teams make this shift, everything changes. Design control stops being the enemy and becomes the ally. Documentation stops being overhead and becomes useful. Compliance stops being a phase and becomes a continuous state.

That's the goal of this framework: not to make compliance easier, but to make it invisible—embedded so deeply in how work gets done that no one thinks about it separately.

Chapter 5 Summary

The regulations never required the constraints that make design control painful. Organizations created those constraints themselves.

FDA, ISO, and IEC standards all allow iterative development. AAMI TIR45 makes this explicit.

Design control, properly implemented, is just good engineering practice made visible. If good engineering is already happening, documentation should be straightforward.

The goal is to shift from 'compliance as overhead' to 'compliance as engineering practice'—and ultimately to make compliance invisible.

PART III — THE FOUNDATION

Before building the framework, we must establish the language and the roles. Precision here prevents confusion everywhere else.

Chapter 6 — The Language That Changes Everything

Words matter. In regulated environments, they matter enormously.

When an engineer says 'risk' and a quality professional says 'risk,' they may not mean the same thing. When someone talks about 'validation,' half the room thinks verification and the other half thinks user testing. When leadership asks about 'compliance status,' they might get a different answer from every person they ask.

This chapter establishes precise terminology that will be used throughout this book and—more importantly—throughout your organization. Disagreement on these terms causes alignment failures that ripple through everything else.

Take the time to internalize these definitions. Print them. Post them. Reference them in reviews, templates, and approvals. Standardize usage across your organization. Precision in language enables precision in execution.

6.1 The ISO 14971 Causal Chain

ISO 14971 defines a specific causal chain that governs all risk reasoning:

HAZARD → HAZARDOUS SITUATION → HARM

This isn't just vocabulary. It's a causal model that determines how you analyze risk, how you design controls, and how you justify residual risk. Getting this wrong leads to risk analyses that look impressive but don't actually protect patients.

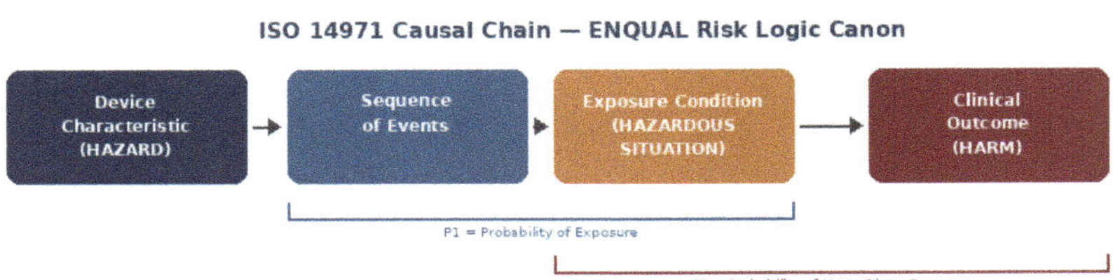

Figure 6-1: ISO 14971 Causal Chain — Device Characteristic to Clinical Outcome with P1/P2 Probability Mapping

FOUR NON-NEGOTIABLE RULES — CAUSAL CHAIN DISCIPLINE

Rule 1: Never call a defect a hazard.

'UI defect', 'software defect', 'design defect' are failure mode labels.

Hazards are device characteristics or information for safety.

Rule 2: Never call user error a hazard.

Use error / incorrect interaction belongs at the Hazardous Situation level.

The hazard is the device characteristic that created the exposure opportunity.

Rule 3: Never skip the Hazardous Situation.

The causal chain is: Hazard → Hazardous Situation → Harm.

Jumping from hazard to harm omits the exposure analysis that determines P1.

Rule 4: Always separate P1 from P2.

P1 = probability of exposure (hazardous situation occurring).

P2 = probability of harm given exposure. ISO 14971 governs P2.

Usability analysis (URRA/IEC 62366-1) refines P1 only.

Post these rules wherever risk files are updated. They prevent the four most common ISO 14971 causal-chain failures seen in FDA inspections.

Figure 6-1 makes the causal chain visible. The hazard is always a device characteristic on the left. User behavior and environmental conditions appear in the middle as the sequence of events that creates the hazardous situation. Harm is the clinical outcome on the right. The two probability dimensions (P1: probability of exposure, P2: probability of harm given exposure) map directly to the P1/P2 framework used in CCB classification (Chapter 9). This chain is the foundation of every risk analysis, risk control decision, and residual risk justification in the framework.

Hazard

A hazard is a potential source of harm. In medical device context, a hazard is always a device characteristic—never a user action, never an environmental condition, never a circumstance. Use-related risk is captured in the hazardous situation (the sequence of events and exposure conditions), not as the hazard itself.

The hazard is something about the device itself that could contribute to harm under certain circumstances.

Example hazards:

- 'Algorithm that produces score without confidence interval'
- 'Display that shows risk score without reference ranges'
- 'Software that permits entry of physiologically impossible values'
- 'System that continues operating when sensor data is stale'

Not hazards (common mistakes):

- 'Clinician misreads the score' — This is user behavior, not a device characteristic
- 'Network failure causes data loss' — This is an environmental condition
- 'User error during data entry' — This is a circumstance, not a device feature

> **REGULATORY ANCHOR:** ISO 14971:2019 defines hazard as 'potential source of harm.' The standard's examples consistently show hazards as device characteristics: 'sharp edges,' 'electrical energy,' 'software malfunction.' User actions are never listed as hazards.

Hazardous Situation
A hazardous situation is a circumstance in which people, property, or environment are exposed to a hazard. This is where user behavior, environmental conditions, and context enter the analysis.

The hazardous situation describes how the hazard manifests in actual use. It connects the device characteristic (hazard) to the potential for harm.

Example hazardous situations:

- 'Clinician views unreferenced score during time-pressured triage decision'

• 'Operator enters value during interruption, doesn't notice implausible result'

• 'System displays stale data during critical assessment, user assumes data is current'

The hazardous situation is where probability enters the analysis. How often does this exposure occur? Under what circumstances? With what frequency?

Harm
Harm is physical injury or damage to the health of people, or damage to property or the environment. This is the actual negative outcome.

Example harms:

• 'Patient receives delayed treatment due to underestimated risk, resulting in adverse outcome'

• 'Misdiagnosis leads to inappropriate therapy'

• 'Patient anxiety from false positive result'

Harm is where severity enters the analysis. How bad is the outcome when it occurs?

> **KEY INSIGHT:** The hazard is always a device characteristic. User behavior appears in the hazardous situation. Confusing these leads to risk controls that train users instead of fixing the device.

> **483 PATTERN:** Risk file lists 'user error' as top hazard with 'training' as primary risk control. Inspector asks: 'If user error is the hazard, why haven't you designed the device to prevent the error?' Organization cannot explain why inherent safety design wasn't the first consideration.

6.2 Terminology Lock

The following terms have specific meanings in this framework. Using them incorrectly creates confusion that propagates through your entire quality system.

⚠️ TERMINOLOGY LOCK — 'User error' as hazard

Correct: User actions appear in hazardous situations, not as hazards. The hazard is the device characteristic that the user interacts with.

SEVERITY RATIONALE PATTERN (use for every risk estimate)

Most likely harm: What is the harm that will occur in the typical case?

Document this with clinical basis or literature support.

Worst plausible harm: What is the worst clinically realistic outcome?

This is not the theoretical maximum — it is the worst outcome that could plausibly occur given the hazardous situation.

Why the severity level chosen is appropriate:

State which harm (most likely or worst plausible) drives your severity classification and why. If you use worst plausible, say so explicitly.
Inspectors will ask: 'Why did you choose that severity level?'
Your rationale must exist in writing at the time of the decision.

Incorrect: Listing 'user error' as a hazard in the risk file

Why it matters: If 'user error' is your hazard, your risk controls will focus on training instead of design. Regulators expect inherent safety design first (ISO 14971 Clause 7.1), then protective measures, then information for safety. Training-only controls suggest you skipped the first two options.

⚠️ TERMINOLOGY LOCK — 'Risk' vs 'Concern'

Correct: Risk = Probability × Severity (quantified or semi-quantified). Risk is a specific combination of likelihood and impact that can be evaluated against acceptance criteria.

Incorrect: Using 'risk' colloquially to mean 'anything we're worried about'

Why it matters: When everything is a 'risk,' risk-based prioritization becomes impossible. Actual risks (with P1/P2/Severity) get buried in lists of concerns. CCB cannot make classification decisions when the word 'risk' means different things to different people.

⚠? TERMINOLOGY LOCK — 'Verification' vs 'Validation'

Correct: Verification = Did we build it right? (against specifications). Validation = Did we build the right thing? (against user needs and intended use).

Incorrect: Using these interchangeably or calling all testing 'validation'

Why it matters: Verification happens throughout development—unit tests, integration tests, system tests against requirements. Validation typically requires a stable release candidate and often involves clinical users or simulated use. Confusing them leads to either premature validation (testing unstable software with users) or missing verification (shipping without proving requirements were met).

⚠? TERMINOLOGY LOCK — 'Engineering-only change'

Correct: No such thing exists in a regulated environment. Every change that affects device behavior requires classification.

Incorrect: Classifying refactoring, performance optimization, or infrastructure changes as outside design control

Why it matters: Any change that affects how the device operates—including how fast it runs, how it handles errors, how it uses memory—is a design change requiring classification. 'Engineering-only' is a compliance evasion pattern that inspectors recognize immediately. The classification might be LOW, but classification must occur.

⚠️ TERMINOLOGY LOCK — 'QA' vs 'RA' vs 'DQE'

Correct: QA = Process compliance and audit (operates at audit cadence). RA = Regulatory strategy and submissions (operates at submission cadence). DQE = Design quality engineering embedded in sprints (operates at sprint cadence).

Incorrect: Using 'QA' to mean anyone who does quality-related work

Why it matters: Conflating these roles creates bottlenecks and accountability gaps. When everyone is 'QA,' no one owns specific decisions. DQE must operate at sprint speed; RA cannot. Asking RA to approve every sprint item creates delays. Asking DQE to make submission decisions creates regulatory risk.

⚠️ TERMINOLOGY LOCK — 'Design Review' vs 'Code Review'

Correct: Design review (per 820.30(e)) is a formal examination of design to evaluate adequacy, identify problems, and propose solutions—typically involving cross-functional perspectives including quality. Code review is a technical review of implementation by engineering peers.

Incorrect: Treating code review as sufficient for design review requirements

Why it matters: Code review verifies implementation quality but doesn't evaluate whether the design is appropriate, safe, or complete. Design review requires quality perspective, risk consideration, and documented outcomes. Many organizations conflate these and have 483 findings for inadequate design review despite having robust code review.

⚠️ TERMINOLOGY LOCK — 'Document' vs 'Record'

Correct: Document = Information that may change (procedures, specifications, plans). Record = Evidence that something happened (test results, approvals, meeting minutes).

Incorrect: Calling everything a 'document'

Why it matters: Documents are controlled by version. Records are controlled by retention. When you update an SRS, you create a new version of a document. When you capture test results, you create a record that shouldn't be changed. Confusing these leads to either over-controlled records (version controlling test results) or under-controlled documents (not managing SRS versions).

⚠? **TERMINOLOGY LOCK — 'SOUP'**

Correct: SOUP (Software of Unknown Provenance) per IEC 62304 includes any software already developed and generally available that is incorporated into a medical device but was not developed under IEC 62304. This includes open-source libraries, commercial SDKs, third-party APIs, and off-the-shelf database engines.

Incorrect: Treating open-source libraries as 'not SOUP because we can see the source code,' or excluding commercial components because they come with documentation.

Why it matters: IEC 62304 requires SOUP identification, functional and performance requirements specification (Class B/C), and known anomaly review. Teams that don't recognize their dependencies as SOUP skip these requirements entirely, creating inspection findings when auditors review the SBOM against the SOUP register. See Appendix AAA for clause-by-class SOUP requirements.

6.3 The Language of Risk Controls

ISO 14971 defines a hierarchy of risk control options:

Table 6-1: ISO 14971 Risk Control Hierarchy

Priority	Control Type	Example	When Appropriate
1st	Inherent safety by design	Eliminate the hazard through design	Always try first

Priority	Control Type	Example	When Appropriate
2nd	Protective measures	Guards, alarms, interlocks	When hazard cannot be eliminated
3rd	Information for safety	Warnings, training, labeling	When above are insufficient

Organizations often jump directly to 'information for safety' (training, warnings) without documenting why inherent safety design or protective measures were rejected. This creates audit findings.

Each control type in the hierarchy maps to specific probability components, and understanding this mapping prevents a common error: selecting controls that feel intuitively satisfying but target the wrong part of the probability chain.

Inherent safety by design primarily reduces Pe or eliminates the hazard entirely. Consider a SaMD application where clinicians enter medication dosages into a free-text numeric field — a common interface element whose hazard is that it accepts any value, including those far outside the safe therapeutic range. If that field is redesigned to present only validated dosage options in a constrained selector, the use error of entering an out-of-range value becomes impossible — Pe drops to zero for that specific error mode. The hazard has been eliminated through design. This is why the standard mandates inherent safety as the first consideration: it addresses root cause rather than compensating for consequences.

Protective measures primarily reduce P0 — they prevent errors that do occur from propagating into hazardous situations. Continuing the dosage example: if the free-text entry field is retained but the system validates entries against the patient's safe dosage range before accepting them, the clinician can still type an incorrect value (Pe is unchanged), but the system intercepts the error before it reaches the patient. Confirmation dialogs before irreversible actions and automated range checks operate at this same P0 level — the error still occurs, but the system catches it before it creates patient exposure.

Information for safety — training, warnings, labeling — attempts to reduce Pe through user education but provides no systemic barrier. In the dosage example, training clinicians to “always double-check the entered value before confirming” depends entirely on human vigilance. If a clinician still enters the wrong value during an interrupted, time-pressured shift, no system mechanism intervenes — neither P0 nor P2 is affected. This is why information for safety occupies the lowest priority: it depends entirely on human performance, which is inherently variable and degrades under stress, fatigue, and interruption — precisely the conditions prevalent in clinical use environments.

This three-tier mapping using the dosage entry example illustrates a principle that applies to every use-related hazard. Appendix FFF.3 explores this mapping in detail with additional examples of risk controls targeting Pe, P0, and P2 respectively.

> **REGULATORY ANCHOR:** ISO 14971:2019 Clause 7.1 requires evaluation of risk control options 'in the following priority order: a) inherent safety by design; b) protective measures in the medical device itself or in the manufacturing process; c) information for safety.' Skipping options requires documented justification.

6.4 P1, P2, and the Severity Matrix

Risk estimation in ISO 14971 involves two probability assessments and severity:

P1: Probability that the hazardous situation occurs (exposure probability)

P2: Probability that the hazardous situation leads to harm (given exposure)

Severity: The magnitude of harm when it occurs

Many organizations conflate P1 and P2, which leads to risk estimates that don't map to reality. Usability findings typically affect P1 (how often users are exposed to the hazard), not P2 (what happens once exposed).

The risk estimation feeds into a risk matrix that determines acceptability. The specific matrix varies by organization and device type, but the structure is consistent: probability (usually combined P1×P2) against severity, yielding a risk level that's compared to acceptance criteria.

Understanding why P1 and P2 are frequently conflated requires recognizing a common mental shortcut: teams ask "how likely is harm?" as a single question, when the standard asks two distinct questions with different data sources and different owners. P1 — the probability that the hazardous situation occurs — depends on device design, use environment, user population, and use patterns. P2 — the probability that the hazardous situation leads to harm — depends on clinical and physiological factors that are largely independent of the device's user interface.

Consider this distinction in practice. A SaMD application displays a risk score without confidence intervals (the hazard). A clinician views this unreferenced score during a time-pressured triage decision (the hazardous situation). The patient receives inappropriate treatment and experiences an adverse outcome (the harm). P1 depends on how often clinicians encounter the unreferenced score during time-pressured decisions — a question answered by use-environment analysis, task analysis, and usability testing. P2 depends on the clinical probability that an inappropriate treatment decision actually produces an adverse outcome — a question answered by clinical literature, epidemiological data, and clinical expert judgment. These are fundamentally different questions requiring different expertise and different data sources.

The conflation problem manifests most visibly in risk matrices. When an organization assigns a single "probability" score that blends P1 and P2 considerations, it becomes impossible to determine which risk controls will actually reduce probability. A risk control that improves the user interface (adding confidence intervals to the display) reduces P1 but has no effect on P2. A risk control that adds clinical decision support (alerting when the recommended treatment conflicts with patient history) reduces P2 but has no effect on P1. Without separating P1 and P2, residual risk estimates after control implementation become arbitrary.

Appendix EEE extends this analysis by decomposing P1 further into Pe (probability of error) and P0 (probability of error causing the hazardous situation), yielding the complete probability chain: Pharm = Pe × P0 × P2. Appendix FFF connects this decomposition to use-related risk analysis required by IEC 62366-1. The decomposition is not merely an academic exercise — it determines where risk controls should be targeted and how residual risk should be estimated.

Chapter 6 Summary

Precise terminology enables precise execution. The ISO 14971 causal chain (Hazard → Hazardous Situation → Harm) determines how you analyze and control risk.

Key distinctions: Hazards are device characteristics, never user actions. Risks are quantified, concerns are not. Verification tests against specifications, validation tests against user needs.

Terminology locks prevent the drift that undermines compliance. Post them. Reference them. Enforce them.

Chapter 7 — Who Does What

One of the most common sources of friction in regulated development is unclear accountability. When everyone is responsible for quality, no one is responsible for specific decisions.

This chapter defines the roles that make the framework function. These aren't job titles—they're accountabilities. One person might hold multiple roles in a small organization, or a role might be shared by a team in a large one. What matters is that each accountability is clearly assigned and that everyone knows who owns what.

7.1 The Core Roles

Product Owner

The Product Owner is accountable for requirements content and prioritization. They own the 'what' of the device.

Owns: Software Requirements Specification (SRS), user needs documentation, intended use definition, feature prioritization

Decision rights: What the device should do, feature priority within regulatory constraints, user workflow design, acceptance criteria

Does not decide: Risk classification (that's DQE), verification adequacy (that's DQE), release readiness (that's a cross-functional determination), regulatory strategy (that's RA)

> **ROLE ACCOUNTABILITY:** Product Owner owns the SRS. When requirements conflict with risk controls or regulatory constraints, Product Owner must work with DQE and RA to resolve—but the Product Owner remains accountable for the final requirements content.

Engineering Lead

The Engineering Lead is accountable for technical implementation and architecture. They own the 'how' of the device.

Owns: Software Design Specification (SDS), code quality, technical architecture, implementation decisions, technical debt management

Decision rights: How requirements are implemented, architecture choices, technology selection, technical tradeoffs, code standards

Does not decide: Risk classification (that's DQE), regulatory strategy (that's RA), release approval (that's Program Manager with DQE/RA input), requirements content (that's Product Owner)

> **ROLE ACCOUNTABILITY:** Engineering Lead owns technical decisions but does not own risk decisions. When engineering proposes a design approach, DQE assesses the risk implications. Engineering Lead cannot self-classify risk.

Design Quality Engineer (DQE)
The DQE is accountable for risk classification and verification adequacy. They're the quality voice embedded in the sprint.

Owns: Risk classification at CCB, verification approval for MEDIUM/HIGH items, traceability validation, risk file maintenance during sprints, SOUP register oversight (IEC 62304 Clause 8), usability-risk integration with UEF, IEC 62304 software safety class determination (Clause 4.3)

Decision rights: Safety impact determination (the CCB classification), rigor level for each change, verification sufficiency, traceability completeness

Does not decide: Feature priority (that's Product Owner), technical approach (that's Engineering Lead), regulatory submission strategy (that's RA), release timing (that's Program Manager)

The DQE role is the key innovation in this framework. Traditional QA operates at audit cadence—checking work after it's done. DQE operates at sprint cadence—participating in work as it happens.

> **ROLE ACCOUNTABILITY:** DQE is the single point of accountability for risk classification. At CCB, DQE doesn't take a vote—DQE makes the determination. Others provide input; DQE decides.

Regulatory Affairs (RA)
RA is accountable for regulatory strategy and submission decisions. They own the relationship with regulatory bodies.

Owns: Submission strategy, regulatory classification, predicate analysis, agency communications, labeling claims, EU Technical Documentation oversight (Annex II), GSPR compliance matrix, Notified Body liaison, PSUR submission for Class III

Decision rights: Whether changes require new submissions, regulatory pathway selection, predicate comparisons, labeling content approval, GSPR compliance determination, Notified Body notification requirements, PCCP boundary determination for AI/ML

Does not decide: Day-to-day risk classification (that's DQE), implementation approach (that's Engineering), sprint-level decisions

RA operates at milestone cadence, not sprint cadence. They should be consulted on HIGH risk items but don't need to approve every sprint item.

> **ROLE ACCOUNTABILITY:** RA is consulted on HIGH risk items and owns submission decisions. RA is not the bottleneck for sprint work—that's DQE's domain. The handoff is clear: DQE handles sprint-level classification, RA handles submission-level strategy.

For EU IVDR devices: RA additionally owns IVDR performance evaluation oversight (Articles 56-58), common specification monitoring (Article 9), and IVD classification determination (Annex VIII). IVDR classification rules differ significantly from MDR — IVD software is typically Class C or D, requiring Notified Body involvement that RA must coordinate.

Project Manager (PM)
The PM is accountable for sprint organization and tracker maintenance. They're the operational coordinator.

Owns: Sprint mechanics, CCB scheduling and facilitation, tracker accuracy, status reporting

Decision rights: Sprint scope commitment, meeting organization, blocker escalation, process improvement

Does not decide: Risk classification (that's DQE), technical approach (that's Engineering), release approval (that's Program Manager)

> **ROLE ACCOUNTABILITY:** PM owns the tracker. When tracker shows BLOCKED items, PM escalates. When CCB queue is long, PM schedules more frequent CCB. PM doesn't make compliance decisions but ensures the system operates smoothly.

Program Manager

The Program Manager is accountable for release decisions and portfolio coordination. They own the overall delivery.

Owns: Release gates, cross-product dependencies, portfolio risk, resource allocation across products

Decision rights: Release timing, residual risk acceptance at portfolio level, cross-team coordination

Does not decide: Individual change classification (that's DQE), sprint-level decisions (that's the sprint team)

> **ROLE ACCOUNTABILITY:** Program Manager is the only role that can approve release. They do so based on evidence that DQE and RA have approved, not by overriding DQE/RA decisions.

PMS Manager

The PMS Manager is accountable for post-market surveillance and field signal analysis. They own the feedback loop from the field.

Owns: Complaint trend analysis, CAPA triggering, field data integration, effectiveness monitoring, EU MDR PMS Plan and PMS Report/PSUR preparation

Decision rights: When field signals warrant design investigation, CAPA initiation, effectiveness criteria

Does not decide: Design changes (those go through CCB like any other change), release decisions

> **ROLE ACCOUNTABILITY:** PMS Manager closes the loop. Field signals that warrant design changes enter Intake and flow through the same Risk Gate as any other change. PMS Manager triggers the signal; the system processes it. For EU-marketed devices, PMS Manager additionally owns the PMS Plan (EU MDR Article 84), PMS Report for Class I (Article 85), and PSUR preparation for Class IIa/IIb/III (Article 86). PSURs for Class III are submitted to the Notified Body through RA.

CCB ACCOUNTABILITY AND BYPASS GOVERNANCE

The DQE leading CCB is accountable for:

- Classification decision rationale — written, dated, and signed
- Traceability expectations — which artifacts must be updated for this tier
- Escalation handling — what triggers escalation to RA or senior DQE
- Exception handling — explicit authorization required, never implied

When CCB is bypassed:

Bypass is a quality system nonconformance requiring investigation and systemic correction — not a coaching moment. A bypassed change that enters a sprint without Safety_Impact classification must be immediately quarantined from sprint completion until classified. The bypass event is documented as a CAPA input. Patterns of bypass trigger temporary reversion to Full Oversight Mode.

7.2 Cadence Separation

Notice that these roles operate at different cadences:

Table 7-1: Role Cadences — How each role operates over time

Role	Primary Cadence	Typical Rhythm
Product Owner	Sprint	Daily involvement in sprint activities
Engineering Lead	Sprint	Daily involvement in sprint activities
DQE	Sprint	Active during sprint, decisions at CCB (weekly)
PM	Sprint	Sprint ceremonies, daily tracker updates
RA	Milestone	Consulted on HIGH items, active at submissions
Program Manager	Release	Release gates, quarterly planning
PMS Manager	Continuous	Ongoing monitoring, periodic reviews

Sprint
Milestone
Release
Continuous

This cadence separation is crucial. The retrofit tax often comes from forcing RA-cadence decisions into every sprint or requiring Program Manager approval for every change.

The framework respects these natural rhythms by involving each role at the right time, not all the time. DQE is embedded in sprints because classification must happen before sprint work begins. RA is consulted on HIGH items because only those might affect submission strategy. Program Manager approves releases, not sprints.

> **KEY INSIGHT:** DQE operates at sprint cadence because risk classification must happen before sprint work begins. RA operates at

milestone cadence because submission decisions require accumulated evidence. Mixing these cadences creates bottlenecks.

7.3 Small Team Adaptations

In small organizations, one person often fills multiple roles. This is acceptable as long as accountabilities remain clear and conflicts of interest are managed.

Common combinations:

• Engineering Lead + Product Owner (one person owns both what and how)

• DQE + RA (one person owns both sprint classification and regulatory strategy)

• PM + Program Manager (one person coordinates both sprints and releases)

Combinations to avoid:

• Engineering Lead + DQE (the person doing implementation shouldn't classify their own risk)

• Product Owner + DQE (the person defining features shouldn't approve their own verification)

The key is that classification and approval functions remain independent from the functions being classified and approved.

> **483 PATTERN:** Same person wrote the code, classified the risk as LOW, and approved verification. Inspector asks: 'Who provides independent review of this engineer's work?' Organization cannot identify anyone.

7.4 The RACI at a Glance

Table 7-2: RACI Matrix — Design Control Activities

Activity	PO	Eng	DQE	RA	PM	PgM
Feature definition	A	C	I	I	I	I
Risk classification	I	C	A	C	R	I
Requirements (SRS)	A	C	R	I	I	I
Risk file updates	I	C	A	C	I	I
Verification approval	I	R	A	I	I	I
Release decision	I	I	R	R	R	A
CCB facilitation	C	C	C	C	A	I
Tracker maintenance	R	R	R	I	A	I
Regulatory submission	I	C	C	A	I	R
Field signal triage	I	I	C	I	I	I

A **Accountable** — *Owns the outcome (single person)*
R **Responsible** — *Does the work*
C **Consulted** — *Provides input before decision*
I **Informed** — *Kept aware after decision*

Table 7-2 shows core sprint activities. For the extended RACI including EU MDR/IVDR activities (GSPR mapping, Technical Documentation, PMS Plan/PSUR, Notified Body audit support), IEC 62304 activities (SOUP register, safety class determination, legacy software gap analysis), and usability engineering activities (UEF maintenance, formative/summative evaluation), see Appendix C.

Chapter 7 Summary

Clear role definition prevents accountability gaps and bottlenecks. Each role has specific decision rights and specific exclusions.

The key innovation is DQE—quality embedded at sprint cadence, not audit cadence. DQE operates fast enough to classify changes before sprints, not slow enough to require retroactive documentation.

Cadence separation respects natural rhythms: DQE at sprint cadence, RA at milestone cadence, Program Manager at release cadence. Mixing these creates the bottlenecks that make compliance feel slow.

Small teams can combine roles, but classification and approval must remain independent from the work being classified and approved.

PART IV — THE FRAMEWORK

Now that you understand the problem and have the vocabulary, here is the solution.

Chapter 8 — The Regulated Sprint

The regulated sprint is the execution engine of this framework. It's not a different type of sprint—it's your existing sprint, viewed through the lens of design control and organized to produce compliance evidence naturally.

Everything in this chapter describes activities your team likely already does. The difference is organization, timing, and accountability. When these align, the retrofit tax disappears.

8.1 The Big Picture

Figure 1 shows the complete regulated sprint process. Study it before reading further—the visual will anchor everything that follows.

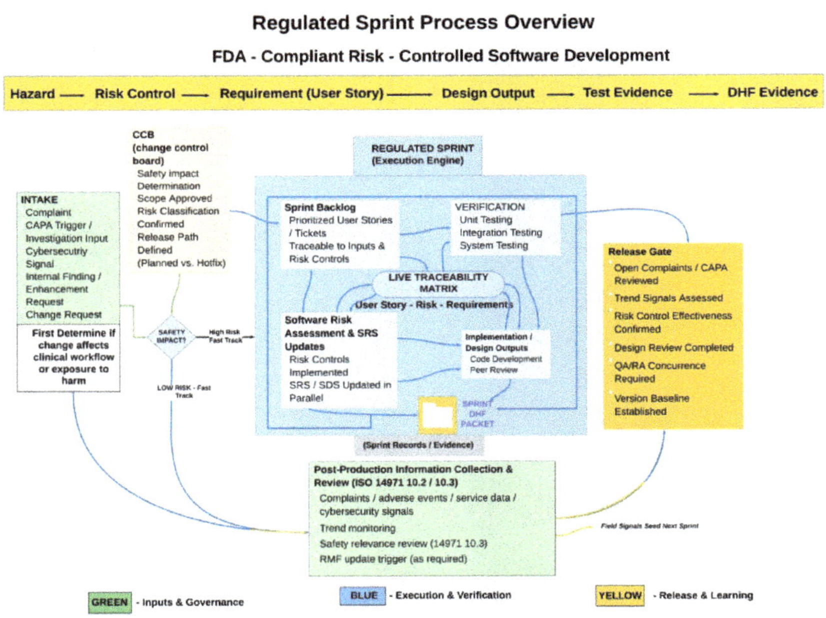

Figure 1: The Regulated Sprint Process Overview

Notice the color coding in the diagram:

GREEN elements are inputs and governance—where changes enter the system and get classified before work begins. This includes Intake (where all signals enter) and CCB (where classification happens).

BLUE elements are execution and verification—where work happens during the sprint. This includes the Sprint Backlog, parallel execution of development and risk activities, and verification.

YELLOW elements are release and learning—where evidence is reviewed and field signals feed back. This includes the Release Gate and Post-Market Surveillance.

8.2 The Loop

The regulated sprint operates as a continuous loop:

> *INTAKE → RISK GATE (CCB) → SPRINT BACKLOG → PARALLEL EXECUTION → VERIFICATION → EVIDENCE → RELEASE GATE → FIELD → PMS → back to INTAKE*

This loop has no endpoint. Each release generates field experience. Field experience generates signals. Signals enter Intake. The cycle continues as long as the device is on the market.

Understanding the loop is crucial: there is no 'done' state. There's only 'current.' The device is always either in active development, deployed in the field, or generating signals that will trigger the next cycle.

8.3 Phase by Phase

Intake

All change signals enter through Intake. This is the single entry point for everything that might affect the device.

Sources of intake signals:

- Product roadmap features (new capabilities)
- Bugs discovered in testing
- Customer complaints

- CAPA actions
- Cybersecurity signals (CVEs, penetration test findings)
- Enhancement requests
- AI model performance drift
- Regulatory changes requiring device updates

The source doesn't matter. What matters is that every signal enters the system through the same door and gets the same classification treatment. No special pathways. No exceptions. No 'urgent' items that skip classification. Emergency and patient-safety signals still route through CCB—the meeting is expedited (hours, not days), but classification happens before sprint entry (see Appendix GG.4).

> **ROLE ACCOUNTABILITY:** Product Owner owns Intake for features and enhancements. PMS Manager owns Intake for field signals. Engineering owns Intake for bugs and technical debt. All items flow to CCB for classification before sprint entry.

Risk Gate (CCB)
Before any item enters a sprint, it must be classified by the Change Control Board. The CCB is the governance mechanism that determines rigor level.

Critical understanding: CCB doesn't approve scope—it classifies risk. The question isn't 'should we do this work?' The question is 'what rigor level does this work require?'

DQE asks the universal question at CCB:

> *Does this change affect clinical workflow or patient safety?*

Based on the answer, the item is classified:

Table 8-1: CCB Classification Criteria and Rigor Levels

Classification	Criteria	Rigor Level
HIGH	Affects clinical decision-making or creates new hazardous situations Causal chain rationale required: Hazard → Hazardous Situation → Harm with P1/P2 basis.	Full rigor: DQE at all stages, RA consulted, dual approval at release
MEDIUM	Affects user workflow but not clinical decisions directly Causal chain rationale required for MEDIUM items touching hazardous situations.	Standard rigor: DQE at key checkpoints, verification approval required
LOW	No clinical impact (infrastructure, performance, cosmetic)	Doc-light rigor: standard code review (no DQE checkpoint); verification evidence captured

HIGH row coloring indicates classification level
MEDIUM row coloring indicates classification level
LOW row coloring indicates classification level

Classification happens before the sprint, not during or after. This is the key enforcement point: no classification, no sprint entry.

> **ROLE ACCOUNTABILITY:** DQE owns risk classification. The classification is not a vote—it's a determination by the accountable role. Others provide input; DQE decides.

> **483 PATTERN:** Items implemented without CCB classification. Inspector asks 'who determined this was low risk?' and no one can answer because classification was skipped or assumed based on the requester's judgment.

Sprint Backlog

Items enter the sprint backlog only after CCB classification. The Safety_Impact field must be populated before items can move to 'In Progress.'

Safety_Impact routes rigor; ISO 14971 evaluates risk in the risk file using Hazard → Hazardous Situation → Harm with P1/P2.

Figure 8-1: Regulated Sprint Board - Sprint 47

Ticket	Title	Safety	CCB Ref	Assignee
BUG-4521	Risk score displays "--" intermittently	MEDIUM	CCB-2026-047	Mike Chen
FEAT-892-2	Alert trigger logic implementation	HIGH	CCB-2026-032	Sarah Lee
FEAT-892-3	Visual alert indicator	HIGH	CCB-2026-032	Sarah Lee
FEAT-892-4	Audio alert tone	HIGH	CCB-2026-032	Tom Park
CHG-104	Dose confirmation dialog	MEDIUM	CCB-2026-041	Mike Chen
CHG-105	Copyright footer update	LOW	CCB-2026-048	Alex Kim

Figure 8-1: Regulated Sprint Board — Sprint 47 showing Safety_Impact classification (a governance triage aligned to risk-based rigor, not the full ISO 14971 risk estimate) for each item

The classification determines the rigor level, which determines which roles must be involved during the sprint:

Table 8-2: Classification Determines Sprint Involvement

Classification	During Sprint	At Verification	At Release
HIGH	DQE involved daily, RA consulted	DQE witnesses critical tests	DQE + RA dual approval + documented benefit-risk conclusion with P1/P2.
MEDIUM	DQE at key checkpoints	DQE reviews and approves evidence	DQE approval
LOW	Standard code review (no DQE checkpoint)	Evidence captured	Included in release packet

HIGH row coloring indicates classification level

MEDIUM row coloring indicates classification level

LOW row coloring indicates classification level

This is how the framework scales: LOW items move quickly with minimal overhead, while HIGH items get the attention they need. The rigor matches the risk.

Parallel Execution

During the sprint, four streams progress simultaneously:

ENGINEERING: Implementation, code review, unit testing, integration

PRODUCT: Requirements refinement as understanding grows, SRS updates

DQE: Risk file updates as design solidifies, verification approach confirmation

TESTING: Test case development, integration testing, system testing

These streams don't wait for each other. They inform each other. When engineering makes a design decision, DQE updates the risk analysis that same day or week. When testing reveals edge cases, Product updates requirements. The documentation evolves with the implementation.

> **KEY INSIGHT:** Parallel execution is not optional. It's how good engineering actually works. The framework just makes it visible and captures it as evidence.

Verification

Verification happens during the sprint, not after. Test execution produces evidence. For MEDIUM and HIGH items, DQE reviews evidence and approves verification adequacy.

The evidence is captured in real time:

- Test results link to test cases
- Test cases link to requirements
- Requirements link to hazards
- DQE approval timestamp shows when verification was confirmed

The traceability exists because it was built during the sprint, not reconstructed later.

Release Gate

At the Release Gate, evidence is reviewed—not created. If the sprint produced complete evidence, the Release Gate is confirmatory. If evidence is missing, the Release Gate surfaces the gap before release.

For releases containing HIGH risk items: DQE + RA dual approval required.

For releases with only MEDIUM or LOW items: DQE approval sufficient.

> **ROLE ACCOUNTABILITY:** Program Manager owns the Release Gate decision. DQE and RA provide approval. The gate confirms that evidence exists—it doesn't evaluate whether work was done correctly (that happened during the sprint).

Field and PMS

After release, the device operates in the field. Field experience generates signals: complaints, usage patterns, performance data, adverse events. PMS Manager monitors these signals and determines when they warrant design investigation.

CLASSIFICATION RATIONALE TEMPLATE (required for MEDIUM and HIGH; recommended for all)

Hazard: [device characteristic / information for safety]

Example: 'Dose entry field accepts values outside therapeutic range without alert'

Hazardous Situation: [exposure circumstance / incorrect interaction]

Example: 'Clinician enters value 10× intended due to ambiguous unit display'

Harm: [clinical consequence]

Example: 'Overdose: hypotension and respiratory depression requiring intervention'

Affected risk control(s): [which existing controls are modified or newly required]

Impacted causal chain ID(s) / Risk_Link(s): [Enter specific HA-IDs from the risk file that this change modifies or adds — e.g., HA-0023, HA-0045. NOT a document title or folder. This is the inspector's direct path from this CCB record to the risk file entry.]

P1 basis: [data/source supporting exposure probability estimate — usability data, literature, field data]

P2 basis: [clinical rationale for harm probability given exposure; risk control effectiveness evidence]

Why this change routes to HIGH / MEDIUM / LOW:

[State the clinical pathway reasoning and what verification rigor is required]

GOVERNANCE NOTE: Safety_Impact routes rigor level. It is not the ISO 14971 risk estimate. The ISO 14971 risk estimate — with P1 (exposure probability) and P2 (harm probability given exposure) — lives in the risk file, not in the tracker field. An inspector who sees Safety_Impact = HIGH will ask: 'Show me the risk file entry.'

When a field signal warrants investigation, it enters Intake—and the loop continues. The complaint becomes a backlog item. The CAPA becomes a design change. The cycle continues.

8.4 The Sprint as Unit of Compliance

A key conceptual shift: the sprint is the smallest auditable unit of design control.

In retrofit organizations, the auditable unit is the release—you prepare a 'DHF packet' at release time by gathering artifacts from various sources.

In this framework, the auditable unit is the sprint. Each sprint produces a Sprint DHF Packet containing all evidence for changes completed in that sprint. Releases are composed of sprints, and the release DHF is composed of sprint DHF packets.

This changes how you think about compliance completeness. Instead of asking 'is the release ready?', you ask 'is this sprint ready?' The smaller question is easier to answer and catches problems earlier.

MOCK INSPECTION Q&A: When Asked About a Specific Change

> **Weak answer:** "Let me find the documents related to that change."
>
> **Strong answer:** "That change was implemented in Sprint 47. Here's the Sprint 47 DHF packet. The CCB record shows classification on March 3rd. The risk file update is dated March 8th. Test evidence was captured March 12th with DQE approval. All timestamps are within the sprint window."

Chapter 8 Summary

The regulated sprint is your existing sprint with explicit design control integration. All signals enter through Intake. All items are classified at CCB before sprint entry. Work and evidence happen in parallel during the sprint.

The sprint is the smallest auditable unit. Sprint DHF packets are produced at sprint close, not at release. Release DHF is composed of sprint packets.

The loop never ends: Intake → CCB → Sprint → Release → Field → PMS → Intake. Understanding the loop is understanding the framework.

Chapter 9 — The Risk Gate in Depth

The Risk Gate—implemented through the Change Control Board (CCB)—is the most important governance mechanism in this framework. Get it right, and everything else flows. Get it wrong, and you've just created a new bottleneck or, worse, a rubber-stamp ceremony that provides no actual governance.

9.1 CCB Is Not What You Think

In many organizations, CCB is an approval meeting. People present their changes, leadership votes, and work proceeds only after approval. This model doesn't scale—it creates the bottleneck that makes teams hate CCB and try to route around it.

In this framework, CCB is a classification meeting. The question isn't 'should we do this work?' The question is 'what rigor level does this work require?'

Work proceeds regardless of classification. A LOW risk item proceeds with reduced documentation burden—but still requires verification evidence and DQE confirmation that reduced rigor is appropriate. A HIGH risk item proceeds with full rigor—parallel risk analysis, DQE involvement throughout, and comprehensive verification.

This distinction is crucial for organizational buy-in. Teams don't resist CCB when it's a 2-minute classification rather than a 20-minute approval negotiation.

9.2 The Universal Question

DQE asks one question of every item:

> *Does this change affect how the device interacts with clinical workflow or patient safety?*

Ticket	Change Description	Classification	DQE Rationale	Key Factor
CHG-101	New drug interaction algorithm	HIGH	Directly affects clinical decision support output	Clinical logic change
CHG-102	Critical alert audio frequency change	HIGH	Affects ability to detect safety-critical notifications	Safety-critical UI
CHG-103	Database query optimization	MEDIUM	No clinical logic change, affects system performance	Reliability impact
CHG-104	Add confirmation dialog before dose	MEDIUM	UI safety enhancement, affects clinical workflow	Workflow change
CHG-105	Update copyright year in footer	LOW	No functional impact, cosmetic only	No safety relation
CHG-106	Fix typo in help documentation	LOW	Non-clinical content, no workflow impact	Documentation only
CHG-107	Change login timeout from 30 to 15 min	MEDIUM	Security change affects clinical workflow continuity	Indirect safety
CHG-108	Add patient photo to verification screen	MEDIUM	Affects patient identification workflow	Identity verification

Figure 9-1: CCB Classification Examples — Safety Impact Determination with DQE Rationale

The question is carefully worded. 'Clinical workflow' captures user-facing changes that affect how clinicians use the device. 'Patient safety' captures changes that could create new hazardous situations or affect existing risk controls.

FIGURE 9-1 READING NOTE — CCB RATIONALE FORMAT

The DQE Rationale column illustrates classification reasoning. In practice, MEDIUM and HIGH rationale entries must use causal chain form:

Hazard: [device characteristic]

Hazardous Situation: [exposure / use error]

Harm: [clinical outcome]

P1 basis: [data/source supporting exposure probability]

P2 basis: [clinical rationale for harm probability given exposure]

Safety_Impact = HIGH/MEDIUM — routes rigor level only; does NOT replace the risk file entry. The risk file carries the complete ISO 14971 analysis.

Why Not "Form, Fit, Function, and Performance"?

Most design teams evaluate changes by asking: "Does this affect form, fit, function, or performance?" That question works for hardware. A machined bracket either fits the mounting hole or it doesn't. A motor either meets torque specifications or it doesn't. The physical properties of the artifact are the risk-relevant properties.

For SaMD, this question collapses. Software has no form. Software has no fit. Every code change "affects function" by definition—that is what code does. The result is a binary gate that cannot discriminate between changing a drug interaction algorithm and updating a copyright year in the footer. Both "affect function." Both get the same answer. And when every change gets the same answer, the question has no classifying power.

This is not a minor distinction. The form-fit-function question is engineering-centric—it asks about properties of the artifact. The universal question is patient-outcome-centric—it asks about the hazardous situation. That reframing maps directly to ISO 14971's causal chain from Chapter 6: hazard → hazardous situation → harm. It is the reason the CCB classification in this framework produces three tiers of proportionate governance instead of a binary pass/fail gate that either rubber-stamps everything or buries every change under full design control rigor regardless of actual risk.

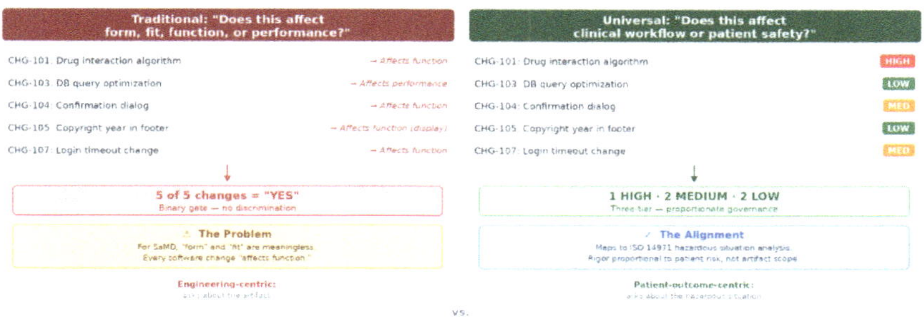

Figure 9-1a: Why "Form, Fit, Function" Fails for SaMD — The Same Five Changes Evaluated by Each Question

Figure 9-1a makes this concrete using the same changes from Figure 9-1. Under the traditional question, five of five changes return "YES"—a binary gate with no classifying power. Under the universal question, those same five changes sort into 1 HIGH, 2 MEDIUM, and 2 LOW—enabling proportionate governance that satisfies both the ISO 14971 mandate for risk-proportional rigor and the development team's need for velocity on low-risk work.

The answer determines classification:

Table 9-1: Answer Patterns to Classification Mapping

Answer Pattern	Classification	Rationale
Yes — directly affects clinical decisions	HIGH	Changes could affect patient outcomes through clinical decision-making
Yes — creates new hazardous situations	HIGH	New hazards require full risk analysis
Partially — affects user workflow	MEDIUM	User experience changes but not clinical logic
Partially — affects display/presentation	MEDIUM	Could affect interpretation but not calculation
No — infrastructure only	LOW	No clinical or user-facing impact
No — performance optimization	LOW	Same behavior, different implementation
No — cosmetic changes	LOW	Visual changes without clinical impact

HIGH row coloring indicates classification level
MEDIUM row coloring indicates classification level
LOW row coloring indicates classification level

> **483 PATTERN:** All items classified as LOW. Inspector reviews sample and finds changes affecting clinical calculations. Asks: 'How was this determined to be low risk?' CCB records show no rationale—classification was automatic or perfunctory.

CRITICAL CONSTRAINT — IEC 62304 Safety Class Floor: CCB classification operates above the floor established by IEC 62304 software safety classification (Clause 4.3). A Class C software system has minimum rigor requirements — including documented architecture, unit testing, and integration testing — that CCB cannot reduce below, regardless of how minor a specific change appears. A 'LOW' classification at CCB for a change within a Class C system still requires the Class C documentation floor. Conversely, CCB may classify an item within a Class A system as HIGH if the change introduces new hazardous situations, requiring more rigor than the base safety class demands. The principle: CCB classification may require MORE rigor than the IEC 62304 class mandates, but never less. See Appendix AAA.3 for the complete interaction matrix.

9.3 What Classification Enables

Classification serves multiple purposes:

For Engineering: It clarifies who needs to be involved and at what points. A LOW item can proceed with standard code review (no DQE checkpoint). A HIGH item means DQE will be reviewing the approach during the sprint, not just at the end.

For QA/DQE: It enables risk-based resource allocation. Not every item needs the same attention. Classification lets DQE focus on items that actually affect safety rather than spreading attention equally across everything.

For Auditors: It demonstrates systematic risk management. The classification record shows that every change was assessed for safety impact before implementation. Even if a particular classification is debatable, the system shows that classification happens consistently.

For Release: It determines gate requirements. If a release contains any HIGH items, RA must approve. If only LOW items, the gate is simpler. Classification enables proportionate governance.

Figure 9-2 makes this concrete. At each classification level, the gating model specifies exactly which design control activities are required (✓), optional (○), or not applicable (—). This is the operational translation of "proportionate governance"—the principle that ISO 14971 and IEC 62304 both mandate: rigor proportional to risk.

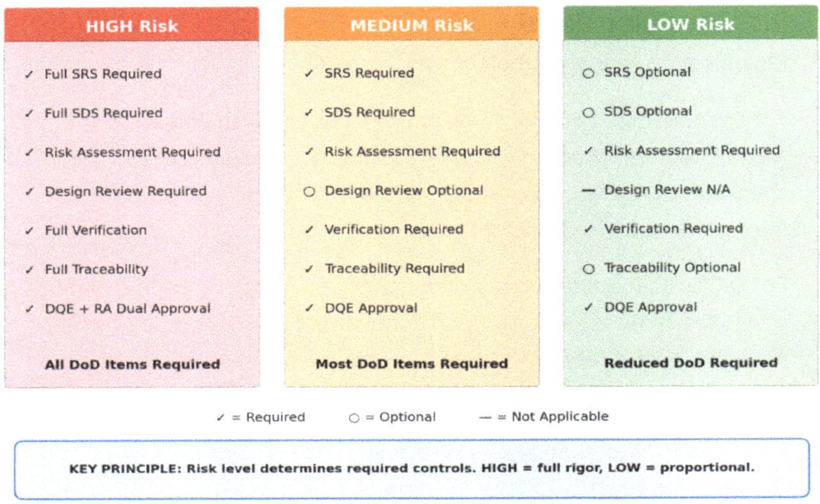

Figure 9-2: Risk-Based Gating Model — Required Controls by Classification Level

Several features of this model deserve attention. First, risk assessment is required at every level—even LOW items require a documented risk assessment to justify the reduced rigor. This is the IEC 62304 floor principle from Section 9.2: CCB may reduce documentation rigor, but never risk awareness. Second, the model distinguishes between "Required" (✓) and "Optional" (○). Optional does not mean "skip it"—it means DQE evaluates whether the specific change warrants it. A MEDIUM change that touches a safety-critical module may still get a formal design review even though the model shows it as optional. Third, the Definition of Done (DoD) requirements scale with risk: HIGH items require all DoD checklist items (SRS, SDS, risk, code review, unit test, integration test, verification, traceability, evidence collection), MEDIUM items require most, and LOW items require a reduced set. This aligns with the three-role sign-off described in Chapter 15—the DoD is how classification translates into sprint-level enforcement.

The critical difference between HIGH and MEDIUM is at the approval gate. HIGH items require DQE + RA dual approval—Regulatory Affairs must concur that the risk controls are adequate and that no submission impact exists. MEDIUM items require DQE approval only, with RA consulted if the DQE determines submission relevance. LOW items still require DQE approval, but DQE is confirming the classification was appropriate and reduced rigor is justified, not reviewing the full evidence chain. For HIGH items, release approval must include a documented benefit-risk conclusion stating most likely harm, worst plausible harm, P1 evidence, and P2 evidence before dual DQE + RA sign-off. The approval signature alone is insufficient; the written conclusion must exist in the release record.

> **KEY INSIGHT:** This gating model is not a permission structure—it is a rigor structure. All three classification levels proceed through the sprint. The difference is how much design control evidence is produced along the way, and who must approve it. Print this figure and post it where CCB meets. It answers the question teams ask most often: "What exactly does a HIGH/MEDIUM/LOW classification mean for my work?"

9.4 CCB Mechanics

A well-functioning CCB meets regularly, processes items quickly, and produces clear records.

Frequency: Weekly is typical for most organizations. Twice weekly if item volume is high. Daily if you're in rapid iteration mode.

Duration: 30 minutes or less for most CCBs. If CCB routinely takes longer, items aren't coming in prepared or classification criteria aren't clear.

Processing time per item: 2-3 minutes for most classifications. Quick items (clearly LOW, similar to previous) take less than a minute. Complex items (genuinely ambiguous, first of a type) might take 5-7 minutes.

The CCB Meeting Flow

1. PM presents the CCB queue (items awaiting classification)
2. For each item, the requester provides a brief description (30 seconds)
3. DQE asks clarifying questions if needed (30 seconds)

4. DQE states classification recommendation and rationale (30 seconds)
5. Engineering provides technical context if DQE requests (30 seconds)
6. DQE confirms or adjusts classification (15 seconds)
7. PM records the decision in the tracker (concurrent)

Total time for a typical CCB with 10 items: 30 minutes or less.

The CCB Record

Each classification decision produces a record containing:

- Item identifier (Jira number or equivalent)
- Item description (what the change is)
- Classification (HIGH/MEDIUM/LOW)
- Rationale (why that classification)
- DQE signature/identifier
- Date

The rationale is crucial. It doesn't need to be lengthy, but it must exist. 'LOW because infrastructure change with no user-facing impact' is sufficient. Empty rationale is not.

For MEDIUM and HIGH items, the required artifact is the CCB Classification Rationale Record in **Appendix NNN.1** — not a summary note. The record must be completed at classification time and stored in the CCB_Reference field before the item enters the sprint queue.

> **REGULATORY ANCHOR:** ISO 14971:2019 Clause 6.2 requires evaluation of identified hazardous situations. The CCB classification is this evaluation in practice—determining whether a change introduces or modifies hazardous situations and what rigor is appropriate.

9.5 Classification Calibration

Classification consistency requires calibration. Without it, different DQEs (or the same DQE on different days) will classify similar items differently.

Calibration approaches:

Reference examples: Maintain a library of classified items that serve as benchmarks. 'This is similar to ITEM-1234, which was classified MEDIUM for the same reasons.'

Periodic review: Monthly review of recent classifications by DQE and RA. Are classifications consistent? Are any clearly too high or too low?

Challenge process: Anyone can challenge a classification. Challenges don't block work but trigger a review. Patterns in challenges reveal calibration issues.

Safety_Impact is a governance routing label — it determines rigor tier, not risk magnitude. An item rated Safety_Impact = HIGH requires the highest documentation and verification rigor. It does not replace the ISO 14971 risk estimate, which requires separate analysis of severity, P1 (exposure probability), and P2 (harm probability given exposure) in the risk management file. These answer two different questions: 'How much process rigor does this change require?' vs 'Is the residual risk from this hazardous situation acceptable?' Both questions must be answered — one by CCB, one by the risk file.

Safety_Impact routes rigor; ISO 14971 evaluates risk in the risk file using Hazard → Hazardous Situation → Harm with P1/P2.

Chapter 9 Summary

CCB is classification, not approval. The question is 'what rigor level?' not 'should we do this?'

The universal question drives classification: 'Does this change affect clinical workflow or patient safety?'

Classification enables risk-based resource allocation—DQE attention goes where it's needed, not everywhere equally.

CCB mechanics matter: weekly, 30 minutes, 2-3 minutes per item, clear records with rationale.

Chapter 10 — Parallel Execution

The traditional model assumes sequential development: requirements, then design, then implementation, then testing. Each phase completes before the next begins.

Iterative development doesn't work that way. Understanding emerges during implementation. Requirements refine as you build. Testing reveals edge cases that affect design. Pretending otherwise—creating documents that say phases are complete when everyone knows they'll change—is the source of much compliance dysfunction.

Parallel execution acknowledges this reality and builds the compliance framework around it.

10.1 What Moves Together

During any sprint, four documentation streams evolve simultaneously:

REQUIREMENTS (SRS): Product Owner refines requirements as engineering reveals implementation implications. Acceptance criteria sharpen. Edge cases are discovered and documented.

RISK (Risk File): DQE updates risk analysis as design decisions expose or mitigate hazards. P1 estimates refine based on design choices. New hazardous situations are identified.

DESIGN (SDS): Engineering documents architecture as it solidifies. Interface specifications firm up. Module boundaries become clear.

TESTING (Test Cases): Test team develops cases as requirements clarify. Test approaches refine based on implementation. Verification coverage expands.

These aren't independent activities that happen to overlap in time. They're interdependent activities that inform each other continuously.

10.2 A Day-by-Day Example

Here's how parallel execution looks in a typical 2-week sprint:

Days 1-2: Sprint Start

Engineering begins implementation based on current requirements understanding. Product Owner has drafted SRS content but knows it may refine.

DQE reviews initial hazard implications based on the design approach. Risk file shows draft analysis with P1/P2 estimates based on current understanding.

Test team begins drafting test cases against current requirements, knowing cases will update as requirements clarify.

Days 3-5: Design Solidifies

Engineering discovers an edge case that wasn't in the original requirement. Product Owner updates SRS to address it. DQE assesses whether the edge case creates new hazardous situations.

Design decision made about error handling. DQE updates risk file with the specific hazard and the design control that addresses it.

Test team adds test cases for the new edge case. Verification approach refines based on design.

Days 6-8: Implementation Completes

Engineering completes implementation. Code review happens. Unit tests pass.

Product Owner confirms SRS accuracy—requirements match what was built.

DQE confirms risk file is current—risk controls match implemented design.

Test team finalizes test cases—coverage matches requirements and risk controls.

Days 9-10: Verification and Close
Test execution happens. Results captured and linked to cases.

DQE reviews evidence, approves verification adequacy.

Sprint ends with all documentation current—SRS, risk file, test evidence all updated and approved.

10.3 The Evidence Pattern

Parallel execution produces a distinctive evidence pattern: document updates distributed throughout the sprint, not clustered at the end.

Healthy timestamp pattern:

- SRS updates: Days 1, 3, 5, 7
- Risk file updates: Days 2, 4, 6, 8
- Test execution: Days 9, 10
- DQE approvals: Days 9, 10

Unhealthy timestamp pattern:

- SRS updates: Day 10
- Risk file updates: Day 10
- Test execution: Day 9
- DQE approvals: Day 10

If all updates happen on the last day, you're still in retrofit mode. The documents were created at the end, not during. Inspectors will notice.

> **483 PATTERN:** All documentation timestamps cluster on final sprint day. Inspector asks: 'If everything was updated on the last day, when were these analyses actually performed?' The timestamps reveal reconstruction.

10.4 Making Parallel Execution Work

Parallel execution requires several enabling conditions:

DQE availability: DQE must be available during sprints, not just at release. If DQE can only review at sprint end, parallel execution is impossible.

Tooling that supports draft states: Documents need to be in draft while work progresses. 'Draft' isn't a quality failure—it's a reflection of appropriate timing.

Cultural acceptance of refinement: Teams must accept that initial versions will change. The SRS on Day 1 won't match the SRS on Day 10, and that's correct behavior.

Communication patterns: Daily standups should include DQE. Design decisions should trigger immediate DQE consultation, not end-of-sprint review.

> **KEY INSIGHT:** Parallel execution isn't a technique. It's a recognition of how work actually happens. The framework just makes it visible and captures the evidence.

CLASSIFICATION-FIRST RULE FOR HIGH CHANGES

For LOW items, parallel execution allows some activities to begin while classification is confirmed — the risk of misclassification is low and quickly corrected.

For HIGH items, risk classification must precede development work. A HIGH-classified change carries the greatest potential for patient impact. Starting implementation before classification is confirmed means development may proceed with the wrong rigor level — verification requirements may be incomplete, risk controls may not be scoped, and DQE engagement may be insufficient.

Governance rule: For any change where the initial assessment suggests HIGH classification, treat it as HIGH until CCB formally confirms otherwise. Do not begin sprint implementation on a potentially HIGH item while awaiting classification. The 30-second delay of a CCB decision costs less than the rework of implementation done at the wrong rigor level.

Chapter 10 Summary

Sequential phases are a fiction in iterative development. Requirements, risk, design, and testing all evolve together.

Parallel execution captures evidence as understanding develops. Timestamps distribute throughout the sprint.

The evidence pattern reveals the truth: distributed timestamps mean concurrent work; clustered timestamps mean reconstruction.

Making parallel execution work requires DQE availability, draft-friendly tooling, and cultural acceptance of refinement.

Chapter 11 — The Design Control Tracker

How do you know if you're ready to release? In most organizations, someone has to go check—pull documents from different systems, verify approvals, confirm traceability. This takes time and introduces risk of missing something.

The Design Control Tracker provides real-time visibility into compliance status. At any moment, anyone can see: which items are complete, which are blocked, and what's preventing release.

Figure 11-1: Bidirectional Traceability - Requirement to Verification

Req ID	Requirement	Design	Hazard	Test	Result
SRS-4.2	Risk score calculation accuracy	SDS-4.2.1	HA-0052	TC-0089	PASS
SRS-9.1	Critical alert threshold config	SDS-9.1.1	HA-0044	TC-0112	PASS
SRS-9.2	Alert trigger within 500ms	SDS-9.2.1	HA-0045	TC-0113	PASS
SRS-9.3	Visual alert prominence	SDS-9.3.1	HA-0046	TC-0114	PENDING
SRS-9.4	Audio alert distinctiveness	SDS-9.4.1	HA-0047	TC-0115	NOT RUN
SRS-5.1	Dose confirmation required	SDS-5.1.1	HA-0015	TC-0078	PASS

Figure 11-1: Bidirectional Traceability — Requirement to Design to Risk to Test

Risk_Link is not 'a link to the risk document.' It is a link to the specific causal chain ID(s) — Hazard → Hazardous Situation → Harm — and the associated risk controls that this change modifies or verifies. A single change can touch multiple causal chains, particularly in SaMD where UI changes often affect multiple use scenarios simultaneously. The Risk_Link field must resolve to a traceable entry in the risk file, not a folder or document title.

11.1 What the Tracker Tracks

For each item in the system, the tracker shows status across multiple dimensions:

Table 11-1: Tracker Field Definitions

Field	Purpose	Owner	Updated
Feature/Change	What the item is	Requester	At creation
Safety_Impact	CCB classification	DQE	At CCB
SRS Status	Requirements documentation	Product Owner	During sprint
SDS Status	Design documentation state	Engineering Lead	During sprint
Risk Status	Risk analysis state	DQE	During sprint
Test Status	Verification evidence	Engineering + DQE	During sprint
Release Status	Overall readiness	Calculated	Automatic
Blockers	What's preventing progress	Anyone	As needed
Evidence Links	Traceability to artifacts	All roles	During sprint

11.2 Status Values

Each status field uses consistent values across the tracker:

NOT STARTED: Work has not begun on this deliverable. Acceptable early in sprint, problematic if still showing late in sprint.

DRAFT: Initial version in progress, not ready for review. Normal state during active sprint work.

UNDER REVIEW: Submitted for cross-functional review or approval. Should not stay in this state for more than a day or two.

APPROVED: Completed and signed off—ready for release. The target state at sprint end.

BLOCKED: Cannot proceed due to dependency, issue, or decision needed. Requires escalation and tracking.

N/A: Not applicable for this change type or risk level. Must be justified by classification.

11.3 Release Readiness Rules

The tracker calculates release readiness based on risk level and status values:

HIGH RISK items: All four columns (SRS, SDS, Risk, Test) must show APPROVED, or N/A with documented justification. Nothing less is acceptable. If any column shows anything other than APPROVED or N/A (with justification), the item is not release-ready.

MEDIUM RISK items: SRS, Risk, and Test must show APPROVED. SDS may be N/A if no architecture impact.

LOW RISK items: Test must show APPROVED. Other columns can be N/A *only when DQE has explicitly confirmed no documentation impact exists*—this is a documented decision, not a bypass.

Any BLOCKED status anywhere: Blocks release until resolved, regardless of risk level.

11.4 The Dashboard View

The tracker dashboard provides executive visibility at a glance:

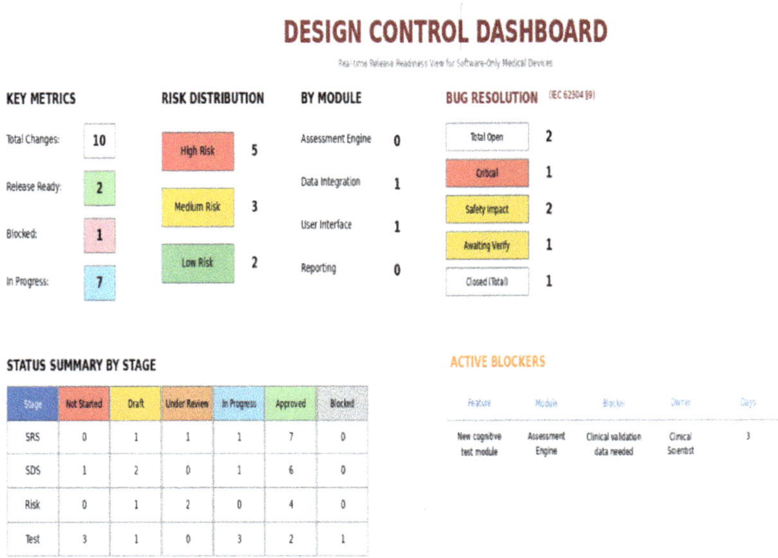

Figure 11-2: Design Control Tracker Dashboard

Key metrics on the dashboard:

- Total changes in the release
- Release-ready count (all required fields APPROVED)
- Blocked count (items that cannot proceed)
- In-progress count (items actively being worked)
- Risk distribution (HIGH/MEDIUM/LOW breakdown)

When leadership asks 'can we release Friday?', the answer is on the dashboard. If everything is GREEN, yes. If there are AMBER or RED items, those specific items need attention.

> **KEY INSIGHT:** The tracker isn't surveillance. It's visibility. When everyone can see the same status, there are no surprises at release time.

11.5 Tracker as Single Source of Truth

The tracker works only if it's the single source of truth. If teams maintain separate tracking spreadsheets, or if tracker status doesn't match actual document status, the tracker becomes useless.

Enforcement mechanisms:

Jira integration: If your tracker is built on Jira, use workflow rules to enforce status updates. Items cannot transition without required fields.

Daily updates: PM verifies tracker accuracy daily. Discrepancies between tracker and reality are escalated immediately.

Release gate check: The Release Gate explicitly verifies that tracker shows GREEN for all items. Discrepancies block release.

Figure 11-3 shows why single-source-of-truth is an architectural claim, not just a policy statement. The Tracker sheet is the master data store. Every other sheet in the system either feeds into it (orange: supporting data linked via ID) or is generated from it (green: formula-driven views). No sheet operates independently. No data is duplicated.

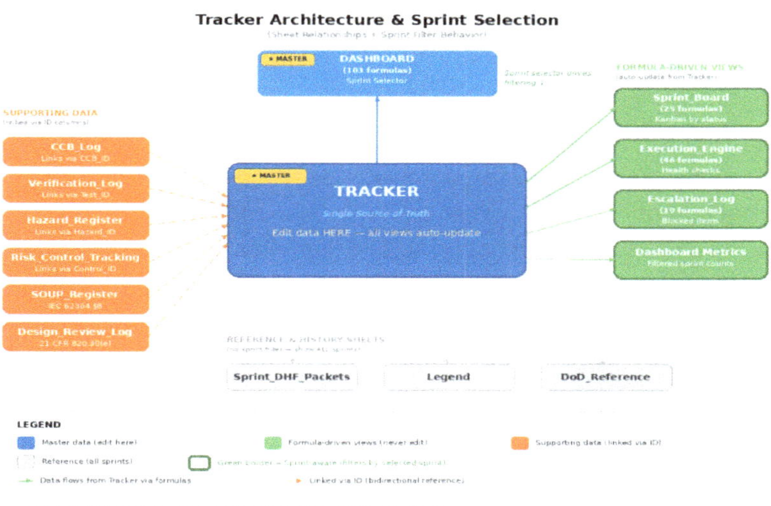

Figure 11-3: Tracker Architecture — Sheet Relationships, Data Flow, and Sprint Selection

The architecture divides into four categories. The blue center is the Tracker itself—the master data store and single source of truth where all item-level information lives. The orange blocks on the left are supporting data sheets (CCB_Log, Verification_Log, Hazard_Register, Risk_Control_Tracking, SOUP_Register, and Design_Review_Log) that hold detail records linked back to the Tracker via Item_ID. When DQE records a CCB decision in the CCB_Log, that decision links to the Tracker row for the same item. When a hazard is identified, the Hazard_Register entry links to the corresponding Tracker item through its Linked_Hazard_ID field. These linkages create the bidirectional traceability that FDA inspectors follow during an audit.

The green blocks on the right are formula-driven views that read from the Tracker but are never edited directly. The Dashboard aggregates Tracker data into release readiness metrics, risk distribution, and gate status—the executive view from Section 11.4. The Sprint_Board presents the same data reorganized by sprint workflow status. The Execution_Engine calculates Definition of Done progress and identifies gaps. The Escalation_Log surfaces blocked items and aging metrics. The Dashboard Metrics section uses COUNTIFS to query Tracker data filtered by the sprint selector. The numbers in parentheses indicate formula count, a measure of how tightly each view is coupled to the master data. The Dashboard alone contains 103 formulas referencing the Tracker.

The figure also shows sprint selection behavior. Sheets with a green border are sprint-aware—they automatically filter based on the Dashboard sprint selector. When the selected sprint changes, all sprint-aware views update simultaneously. This centralized control eliminates the "which sprint am I looking at?" confusion during sprint planning, release readiness reviews, and hotfix tracking. The gray blocks at the bottom (Sprint_DHF_Packets, Legend, DoD_Reference) are reference and history sheets that always show all sprints—they constitute the fourth category.

This architecture has a direct regulatory implication: when an FDA inspector asks "show me the risk analysis for this change," you do not search across disconnected documents. You navigate from the Tracker row to the Hazard_Register entry to the Risk_Control_Tracking record—all linked by ID, all formula-verified for consistency. This is what QMSR means by a “process-based quality management system”: interconnected records that demonstrate systematic control, not isolated documents that happen to describe the same device. A comprehensive reference implementation of this tracker is provided in Appendix GGG.

> **ROLE ACCOUNTABILITY:** PM owns tracker accuracy. When tracker shows APPROVED but the actual document isn't approved, PM escalates and fixes. Tracker status must match reality.

> **483 PATTERN:** Tracker shows all items APPROVED, but inspector asks to see the approval signature on a specific document—and it doesn't exist. Tracker status was updated without verifying actual approval.

Under QMSR and CP 7382.850 (effective February 2, 2026), FDA inspectors will increasingly ask process-flow questions: “Show me how a design change moved through your system.” A well-maintained tracker provides the answer with timestamped evidence—traceability from risk classification through verification to DHF approval. The tracker doesn’t just document compliance; it demonstrates process effectiveness through linkages that can be followed in real-time during an inspection.

RISK_LINK TRACEABILITY RULES

Risk_Link = a pointer to a specific causal chain ID in the risk file.

Not: a folder name, document title, or risk register tab.

Required content at the linked destination:

Hazard → Hazardous Situation → Harm

P1/P2 probability estimates for the chain, associated risk controls, current residual risk status.

One change can touch multiple causal chains.

This is especially common in SaMD where a UI change may affect multiple use scenarios simultaneously, or where an algorithm change touches both functional performance and display presentation hazards.

When a change affects multiple chains, Risk_Link must point to all of them.

A single-chain Risk_Link on a multi-chain change is an incomplete record.

Inspector question: 'For this change, show me every causal chain affected and confirm each chain's risk controls were re-evaluated.'

Answer: pull the Risk_Link entries. If they don't cover all affected chains, that is the finding.

Chapter 11 Summary

The tracker provides real-time visibility into compliance status. Every item, every status dimension, every blocker—visible at a glance.

Release readiness is calculated automatically based on risk level and status values. No manual assessment needed.

The tracker must be the single source of truth. Status must match reality. Daily verification by PM ensures accuracy.

Chapter 12 — Integrating Your Tools

Most engineering teams already use Jira or similar tools for work tracking. The good news: you don't need to replace your tools. You need to extend them.

This chapter shows how to configure your existing tools to support the framework. The specific examples use Jira, but the principles apply to any work tracking system.

12.1 The Core Custom Fields

Add these custom fields to your Jira configuration:

Table 12-1: Jira Custom Field Configuration

Field Name	Type	Purpose	Set By	When
Safety_Impact	Dropdown (HIGH/ MEDIUM/LOW)	CCB classification	DQE	At CCB, before sprint
CCB_Reference	Text/Link	Link to CCB record	PM	At CCB
SRS_Link	URL	Traceability to requirements	Product Owner	During sprint
Risk_Link	URL	Traceability to risk file	DQE	During sprint
Verification_Status	Dropdown	DQE approval of evidence	DQE	At verification

These five fields transform your Jira ticket from an engineering work item into a design history record.

12.2 Workflow Enforcement

Configure Jira workflow rules to enforce the process:

Rule 1: Items cannot move to 'In Progress' without Safety_Impact value. This enforces CCB classification before work begins. No exceptions.

Safety_Impact routes rigor; ISO 14971 evaluates risk in the risk file using Hazard → Hazardous Situation → Harm with P1/P2.

Rule 2: Items cannot move to 'Done' without Verification_Status for MEDIUM/HIGH items. This enforces DQE approval before completion.

Rule 3: Items cannot move out of 'CCB Queue' status without CCB_Reference. This enforces record creation at CCB.

These rules aren't bureaucracy—they're guardrails. They prevent items from slipping through without classification, which prevents the '483 pattern' of unclassified changes.

12.3 Jira as Design History File

Here's the key insight: your Jira ticket is already a design history record. It contains:

What was requested: The story/bug description captures the design input.

What was decided: Comments, acceptance criteria, and linked decisions capture design rationale.

What was implemented: Linked commits, PRs, and branches capture the design output.

What was tested: Linked test executions capture verification evidence.

The custom fields make this record regulatory-complete by adding classification, traceability, and approval status.

> **KEY INSIGHT:** You don't need a separate compliance system. You need to make your existing engineering system compliant. The Jira ticket becomes the design history record.

REGULATORY NOTE: When Jira serves as the design history record system with electronic approvals and signatures, the system must comply with 21 CFR Part 11 (Electronic Records; Electronic Signatures) and, for EU-marketed devices, EU Annex 11 (Computerised Systems). This includes audit trail requirements, access controls, and signature authenticity. See Appendix N for tool validation requirements and Part 11 compliance considerations.

12.4 Traceability Through Links

Traceability in this framework is achieved through links, not matrices maintained in separate documents.

Each Jira item links to:

- SRS section (via SRS_Link field)
- Risk file entry (via Risk_Link field)
- Test execution results (via Jira test management integration)
- Code commits (via Jira-Git integration)
- CCB record (via CCB_Reference field)

The traceability matrix isn't a document—it's a query. Run a report showing items, their links, and their status. The matrix generates itself from the link data.

12.5 Sprint DHF Packet Generation

At sprint end, the Sprint DHF Packet is generated from Jira data:

1. Query all items completed in the sprint
2. For each item, export: description, classification, links, approvals, timestamps
3. Attach or link the SRS/Risk/Test documents at their sprint-end versions
4. Generate traceability report from link data
5. Package into Sprint DHF folder

This can be automated or manual, but the data source is always Jira. The Sprint DHF Packet is a view of Jira data, not a separately maintained document.

> **483 PATTERN:** DHF documents don't match Jira records. Inspector compares the 'official' DHF to Jira and finds discrepancies. Which one is correct? Neither can be trusted.

VERIFICATION ≠ VALIDATION — AN INSPECTION-CRITICAL DISTINCTION

Verification: Did we build the system correctly?

Answers to design specifications (SRS). Tests confirm the output matches the specification. Verification evidence belongs in the sprint DHF packet and is linked from the Jira item via Verification_Status and test links.

Validation: Did we build the correct system?

Answers to user needs and intended use. Validation confirms the device as a whole performs as intended in the use environment. Validation is not sprint-level — it occurs at the system or release level.

Common SaMD failure: calling usability testing 'validation' when it is actually formative evaluation (a verification activity against design specifications). Summative evaluation IS validation — it must be performed against the final production-equivalent device and use environment.

For each verification test, the record should answer:

'Which design specification does this test verify, and what acceptance criteria confirm that the relevant risk control is effective?'

Without the risk control linkage, verification is technically complete but inspection-incomplete — the auditor cannot trace from test result to risk file.

Chapter 12 Summary

Five custom fields transform Jira into a design history system: Safety_Impact, CCB_Reference, SRS_Link, Risk_Link, Verification_Status.

Workflow rules enforce the process: no work without classification, no completion without verification approval.

Traceability is achieved through links, not separate matrices. The matrix is a query result, not a maintained document.

Sprint DHF Packets generate from Jira data. One source of truth, multiple views.

PART V — MAKING IT WORK

Theory is valuable, but implementation is what matters. This section addresses the practical challenges: integrating usability engineering with risk management, building a post-market surveillance loop that satisfies both FDA and EU MDR requirements, defining release gates that prevent incomplete evidence from shipping, and handling the emerging signal types—AI/ML performance drift, cybersecurity vulnerabilities, and continuous learning systems—that are reshaping regulated software.

Chapter 13 — Usability and Use-Related Risk

Usability is one of the most misunderstood elements of design control. Many organizations treat it as a parallel track—something the UX team does separately from risk management.

In this framework, usability feeds directly into risk analysis. Usability findings don't create new hazards—they reveal hazardous situations that already exist in actual use.

13.1 The Integration Point

Remember the causal chain: Hazard → Hazardous Situation → Harm

The hazard is a device characteristic. When usability testing reveals that users struggle with an interface, that struggle is evidence about the hazardous situation—a circumstance where the user is exposed to the hazard.

Example integration:

- Hazard (device characteristic): Display shows score without reference ranges
- Usability finding: In testing, 3 of 10 users misinterpreted scores
- Hazardous situation refined: Users misinterpret unreferenced scores during time-pressured clinical decisions
- Risk analysis updated: P1 estimate (probability of hazardous situation) refined based on observed data
- Risk control decision: Add reference ranges to display (design change, not training)

The design principles for implementing usability engineering in SaMD are detailed in AAMI HE75:2025, which provides practical 'how-to' guidance for achieving IEC 62366-1 requirements. HE75:2025 Section 5 establishes the conceptual framework for managing use error risk, Section 9 provides updated usability testing guidance including remote evaluation methods particularly relevant for SaMD, and Section 21 addresses software user interface design principles for clinical displays and decision support. See Appendices XX and LLL for HE75:2025 application within this framework.

13.2 Formative vs. Summative

IEC 62366-1 distinguishes between formative and summative usability activities:

Formative evaluation happens during development. It identifies use problems while design can still change easily. Findings flow into the sprint like any other design input.

Summative evaluation happens on the release candidate. It demonstrates that use-related risk has been reduced to acceptable levels. Findings may trigger additional sprints if issues are severe.

Both types of findings update the risk file. Formative findings might change P1 estimates or identify new hazardous situations. Summative findings provide verification evidence that risk controls are effective.

> **REGULATORY ANCHOR**: IEC 62366-1:2015/AMD1:2020 Clause 5.9 requires that results of summative evaluation be documented in the usability engineering file and integrated with the risk management file. The standard explicitly requires integration, not parallel tracking.

13.3 What Usability Testing Produces

Usability testing produces evidence that updates the risk file:

P1 refinement: Observed user behavior provides data to estimate how often hazardous situations actually occur. If 3 of 10 users struggle, that's data for P1.

RELEASE DECISION INTEGRITY — HIGH RISK CHANGES

For any HIGH-classified change, the release decision must document:

1. **Benefit-risk conclusion:** Is the residual risk acceptable in relation to benefits?

   This is a documented determination, not an implied one.

2. **Severity rationale:**

   Most likely harm: [clinical outcome in the typical exposure case]

   Worst plausible harm: [worst clinically realistic outcome]

   Severity level used: [which harm drives the classification and why]

3. **P1 addressed:** What evidence supports the exposure probability estimate? (Usability data, literature, field data — cited specifically.)

4. **P2 addressed:** What evidence supports the harm probability given exposure? (Clinical rationale, risk control effectiveness evidence — cited specifically.)

AVOID: 'Low probability makes high severity acceptable' without stating this explicitly in benefit-risk terms. An inspector who sees a HIGH-severity item released will ask: 'Show me the benefit-risk reasoning.' It must be in writing.

New hazardous situations: Testing may reveal interactions between users and the device that weren't anticipated. These are new hazardous situations requiring analysis.

Risk control verification: If a risk control was implemented to address a use-related hazard, usability testing can verify whether the control is effective.

This evidence enters through the same process as any other signal: if usability testing reveals issues requiring design changes, those changes enter Intake → CCB → Sprint.

ROLE ACCOUNTABILITY: DQE owns the integration between usability findings and the risk file. When the usability team reports findings, DQE assesses impact on hazardous situations and updates risk documentation.

483 PATTERN: Usability findings documented but not integrated with risk file. Inspector finds usability report showing user struggles, but risk file shows no corresponding hazardous situations or P1 updates. The two systems are disconnected.

URRA boundary (governance rule): Use-related analysis refines P1 — the probability of exposure — by improving understanding of interaction pathways and hazardous situation occurrence. It does not redefine hazards (which remain device characteristics), does not estimate P2 (which ISO 14971 governs), and does not replace ISO 14971 risk analysis. IEC 62366-1 and ISO 14971 are parallel processes with defined integration points — neither subsumes the other.

Chapter 13 Summary

Usability is not parallel to risk management—it feeds into risk management. User behavior appears in hazardous situations, not as hazards.

Formative findings flow into sprints. Summative findings verify risk control effectiveness.

Usability data refines P1 estimates, identifies new hazardous situations, and verifies risk controls. All of this updates the risk file.

Chapter 14 — Post-Market Surveillance

Post-market surveillance isn't a separate activity from design control—it's the field data source that feeds back into the loop. The device in the field generates signals. Those signals, properly processed, become design inputs for future changes.

14.1 Complaints as Design Input

When a customer reports a problem, that report is evidence of a potential hazardous situation occurring in the real world. It's not just a customer service issue—it's design input.

The complaint tells you: a hazardous situation that you analyzed theoretically has now occurred in practice. Your P1 estimate (probability of the hazardous situation) now has real-world data.

The PMS loop for complaints:

1. Complaint received by customer support
2. Complaints Specialist triages: Is this a complaint per regulatory definition?
3. If yes, logged in complaint system with required data

AI/ML CAUSAL CHAIN EXAMPLE — MODEL DRIFT

AI and machine learning changes are device characteristics. They follow the same ISO 14971 causal chain discipline as any other hazard.

Hazard: Model performance degradation due to data distribution shift (device characteristic: the algorithm's sensitivity to training data distribution)

Hazardous Situation: Clinician is exposed to model output that no longer reflects current population — recommendation presented with unwarranted confidence

Harm: Missed diagnosis or delayed therapy based on incorrect model output

Triggers requiring risk re-evaluation:

- Changes to model logic, weights, or architecture
- Changes to training data sources or preprocessing
- Changes to deployment environment or intended patient population
- Changes to security posture or authentication (threat environment = hazard)
- Performance drift signals from field monitoring

P1 basis: Exposure occurs whenever the model is queried against a patient population that has shifted from training distribution; frequency depends on deployment scope and monitoring interval. URRA characterizes clinician interaction pathways but does not replace this exposure estimate.

P2 basis: P2 is driven by whether the clinician acts on a degraded recommendation before harm occurs. Clinical literature supports elevated harm probability where outputs are presented with high confidence and no performance indicator — clinicians have no signal to question the recommendation. P2 is partially mitigated only if the device displays a confidence level or out-of-distribution warning at the point of use.

URRA boundary: Use-related analysis refines P1 (exposure to the hazardous situation) by characterizing how clinicians interact with model outputs. It does not redefine the hazard (model characteristic) and does not replace ISO 14971 risk analysis.

4. PMS Manager reviews for trends and safety significance
5. If trend detected or safety concern identified, signal enters Intake
6. CCB classifies the signal like any other design input
7. If design change warranted, enters sprint
8. Fix deployed, effectiveness monitored
9. CAPA closed when effectiveness confirmed

14.2 Trend Analysis

Individual complaints may not trigger action, but trends might. PMS Manager is responsible for identifying trends that warrant investigation.

Trend triggers (examples):

- Same failure mode reported by multiple users
- Increasing frequency of a complaint type
- Complaints correlating with specific software version or configuration
- Complaints affecting specific user populations

When a trend is identified, PMS Manager documents the analysis and enters a signal into Intake. The signal might result in:

- Design investigation (if root cause is unknown)
- Design change (if root cause points to device)
- Training or labeling update (if root cause is user-side but addressable)
- Monitoring continuation (if trend is inconclusive)

> **ROLE ACCOUNTABILITY:** PMS Manager owns trend analysis and CAPA triggering. When a CAPA requires design changes, those changes flow through CCB like any other item. PMS Manager doesn't bypass the process—they feed it.

14.3 Closing the Loop

A closed loop means you can trace from field signal back to design response:

This complaint → triggered this CAPA → resulted in this design change → verified by this test → deployed in this release → no recurrence in subsequent complaints

If you can't trace this chain, your loop isn't closed. Somewhere there's a gap between the field signal and the design response.

INSPECTION Q&A: CAUSAL CHAIN AND P1/P2 RIGOR

Inspector: 'Show me the hazard, hazardous situation, and harm for this change.'

Strong answer: Produce the risk file entry. Point to: the device characteristic (hazard), the exposure event or use error pathway (hazardous situation), and the clinical outcome (harm). If the entry calls use error a hazard, or skips the hazardous situation, stop and acknowledge the gap — do not try to explain it away. The fix is a risk file update with a CAPA.

Inspector: 'Show me how you estimated P1 and P2, and what objective evidence supports each.'

Strong answer: 'P1 — exposure probability — was estimated from [usability data / literature / field data]. IEC 62366-1 URRA informed this estimate; it did not replace our ISO 14971 risk analysis. P2 — harm probability given exposure — was estimated from [clinical rationale / literature / analogous device data]. These are in our risk file at [specific section/ID]. Here are the timestamps showing this analysis was completed before the change was released.'

If you cannot produce separate P1 and P2 rationale with supporting evidence, that is the finding. Not the missing paperwork — the missing analysis.

Closing the loop also means updating the risk file. If a complaint reveals that a hazardous situation occurred, the risk file should reflect that. P1 estimates based on theoretical analysis should be updated with field data.

> **REGULATORY ANCHOR:** ISO 14971:2019 Clause 10.2 requires that information from production and post-production be collected and reviewed. Clause 10.3 requires that this information be evaluated for relevance to safety. The framework implements these requirements through the PMS → Intake → CCB pathway.

> **483 PATTERN:** Complaints logged but not connected to risk file or design changes. Inspector traces a complaint and finds no evidence that it was evaluated for safety significance or that the risk file was updated. The loop is not closed.

14.4 EU MDR Post-Market Surveillance Obligations

For devices marketed in the EU, post-market surveillance requirements under EU MDR 2017/745 are substantially more prescriptive than FDA requirements and must be addressed alongside the complaint-driven PMS described above.

EU MDR Article 83 mandates a documented PMS system for each device, suited to actively collecting and using relevant data — the framework's PMS loop with field signals entering Intake satisfies the 'active collection' requirement. Article 84 requires a formal PMS Plan covering systematic data collection procedures, methods for managing incidents and field safety corrective actions, and a PMCF plan where applicable. Article 85 requires a PMS Report for Class I devices summarizing results and conclusions. Article 86 requires Periodic Safety Update Reports (PSURs) for Class IIa/IIb/III devices, updated at least annually for Class IIb/III and at least every two years for Class IIa, with Class III PSURs submitted to the Notified Body.

The framework's PMS loop, trend analysis, and CAPA integration (Sections 14.1-14.3) satisfy the operational requirements of Articles 83-86. However, organizations must ensure the EU-specific documentation artifacts — PMS Plan, PMS Report or PSUR, and PMCF Plan where applicable — are produced and maintained. See Appendix LLL for comprehensive EU MDR/IVDR PMS alignment.

ROLE ACCOUNTABILITY: PMS Manager owns EU MDR PMS documentation: PMS Plan (Article 84), PMS Report for Class I (Article 85), and PSUR coordination for Class IIa/IIb/III (Article 86). RA owns the Notified Body submission of PSURs for Class III devices. DQE provides input on risk file implications from PMS findings. This division mirrors the existing PMS accountability structure — PMS Manager owns field data analysis, RA owns regulatory submissions.

Post-market signals must be normalized into the existing causal chain structure (Hazard → Hazardous Situation → Harm). A complaint describing a use error maps to the Hazardous Situation level — not the Hazard level. If a signal maps to an existing chain, update P1 or P2 estimates as warranted by the new field evidence. If no existing chain covers the signal — a new use pathway or device characteristic is implicated — a new causal chain must be added to the risk file before the signal can be closed. PMS signals cannot be resolved in a parallel taxonomy; every signal must resolve into the risk file's causal chain structure.

Causal discipline reminder: Environment contributes to hazardous situation; hazard remains the device characteristic (including data handling, logic, and availability design choices). When a post-market signal points to network failure, integration issues, or clinical workflow conditions, these are environmental contributors to a hazardous situation — not hazards themselves. Trace them to the device characteristic at fault before updating the risk file.

COMPLAINT MAPPING EXAMPLE — CAUSAL CHAIN NORMALIZATION

Complaint received: 'Device displayed wrong unit (mL vs cc); clinician set rate 10× intended. Patient received overdose. MDR filed.'

Step 1 — Map to causal chain:

Hazard (chain ID): Dose display labeling characteristic (unit presentation)

Hazardous Situation: Clinician misreads unit, sets 10× intended rate

Harm: Overdose, hypotension, medical intervention required

Step 2 — Assess P1 impact: Did this complaint reveal that exposure is more frequent than previously estimated?

> *Demonstrated:* Yes — this is an MDR event, meaning the hazardous situation (unit misread → 10× overdose) actually occurred in the field. Prior P1 estimate must be updated upward. If the risk file estimated P1 as "rare" or based only on usability study data, this complaint is new objective evidence requiring a revised P1 entry with the complaint number as source.

Step 3 — Assess P2 / severity: Did the harm severity align with worst plausible harm in the risk file? If severity was underestimated, update the risk file.

> *Demonstrated:* Harm reported (hypotension, medical intervention required) aligns with the expected worst plausible harm for a 10× overdose. P2 basis: once the clinician set the incorrect rate, no design control intercepted the administration before harm occurred — P2 is high. If the risk file entry lacked a design control that reduces P2 (e.g., hard-stop validation at point of entry), this complaint is evidence that the risk control gap was real and must be addressed in CAPA.

Step 4 — Resulting action: CAPA (risk control update) + RMF update + IEC 62366-1 use scenario re-evaluation if a new use error pathway was identified.

Closing a complaint without completing Steps 1-4 is a quality system gap.

Chapter 14 Summary

Complaints are evidence of hazardous situations in the field. They're design input, not just customer service issues.

Trend analysis by PMS Manager identifies signals that warrant design investigation. Signals enter Intake and flow through CCB.

Closing the loop means tracing from complaint to CAPA to design change to verification to deployment to effectiveness confirmation.

For EU-marketed devices, EU MDR Articles 83-86 impose additional PMS documentation obligations: PMS Plan, PMS Report (Class I), and PSURs (Class IIa/IIb/III). PMS Manager owns these artifacts; RA owns Notified Body PSUR submissions for Class III.

Chapter 15 — Release Gates and Definition of Done

Two concepts are crucial to making the framework work: the sprint-level Definition of Done and the release-level Release Gate. Together, they ensure nothing ships without complete evidence.

15.1 Definition of Done (Sprint Level)

The regulated Definition of Done requires sign-off from three perspectives:

ENGINEERING confirms: Code complete, peer-reviewed, unit tests passing, integration tests passing

PRODUCT confirms: Requirements linkage confirmed, SRS current, acceptance criteria met

DQE confirms: Verification evidence approved, traceability complete, risk file current (for MEDIUM/HIGH items)

An item isn't Done until all three sign-offs exist. This is enforced by Jira workflow: items cannot move to Done without Verification_Status = APPROVED for MEDIUM/HIGH items.

The three-role sign-off prevents any single perspective from declaring completeness. Engineering might think code is done, but if DQE hasn't approved verification, it's not Done. Product might think requirements are met, but if test evidence doesn't exist, it's not Done.

15.2 Release Gate (Release Level)

The Release Gate confirms that evidence is complete for all items in the release. It's confirmatory, not evaluative—the evaluation happened during sprints.

Release Gate checklist:

- All items have Safety_Impact classification (no TBD or blank)
- All HIGH items have DQE + RA approval
- All MEDIUM items have DQE approval
- All evidence links are populated
- Tracker shows no BLOCKED items

- No unresolved complaints affecting this release
- Version baseline established

Note on risk controls: Training-only controls are not sufficient as the primary risk control for MED or HIGH items unless the risk file documents why inherent safety design and protective measures were not applicable or practicable (ISO 14971 Clause 7.1 hierarchy). If the release gate reveals a MED/HIGH item whose only listed control is training or labeling, this is a gate-blocking finding.

If everything in the sprint was Done correctly, the Release Gate takes minutes. If gaps exist, the Release Gate surfaces them—better to find gaps here than after deployment.

Figure 15-1 shows what this looks like in practice. The Release Gate Dashboard for Sprint 47 displays both Evidence Status and Gate Status at a glance—the PM can immediately see whether all items are green or whether gaps require resolution before release.

The governance definition of success: your organization moves at speed while maintaining stable, contemporaneous evidence; causal discipline is preserved in every risk file entry (Hazard → Hazardous Situation → Harm, with P1 and P2 separately addressed); risk controls are verified before release; PMS signals close the loop back into the risk management process. Compliance is not a separate activity — it is the natural output of engineering done correctly.

Figure 15-1: Release Gate Dashboard — Sprint 47 showing Evidence Status and Gate Status

15.3 When Release Gate Fails

If the Release Gate reveals gaps, the response is:

For missing evidence: Determine if evidence exists but isn't linked (fix immediately) or doesn't exist (requires rework)

For incomplete approvals: Determine if approval is pending (expedite) or work isn't done (block release)

For BLOCKED items: Determine if block can be resolved before release or if item must be removed from release

Release Gate failure should be rare if sprints are executed correctly. If Release Gate frequently fails, the problem is sprint execution, not the gate itself.

> **ROLE ACCOUNTABILITY:** Program Manager owns the Release Gate decision. They can release only when all conditions are met. They cannot override DQE or RA decisions—they confirm those decisions exist.

KEY INSIGHT: Definition of Done is sprint-level enforcement. Release Gate is release-level confirmation. Together they ensure nothing ships without complete evidence.

15.4 The Validation Sprint

Some validation activities cannot be completed within a normal sprint cadence. Summative usability testing, extended reliability testing, and clinical validation may require a dedicated Validation Sprint.

The Validation Sprint differs from regular sprints:

- Code freeze on the Release Candidate (no new development)
- Focus on validation activities only
- Any issues found trigger new design changes (which restart the cycle)
- Release Gate occurs only after successful validation

The key discipline: development cannot continue on the validated code while validation is in progress. Changes invalidate the validation. Development continues on a branch for the next release.

For EU-marketed devices, clinical evaluation (EU MDR Article 61) or performance evaluation (EU IVDR Articles 56-58) may be required as part of validation. These activities should be planned alongside the Validation Sprint, with results feeding into the Technical Documentation and informing the benefit-risk determination required by Annex I GSPR. See Appendix LLL for EU-specific validation requirements.

Chapter 15 Summary

Three sign-offs required at sprint level: Engineering, Product, DQE. An item isn't Done until all three perspectives confirm completeness.

The Release Gate is confirmatory, not evaluative — it verifies that sprint-level evidence is complete. If sprints execute correctly, the gate takes minutes.

Program Manager owns the Release Gate decision but cannot override DQE or RA approvals. Validation Sprints freeze code while validation completes; for EU-marketed devices, clinical evaluation may be required alongside validation.

Chapter 16 — AI, Cybersecurity, and Emerging Signals

New signal types are emerging: AI model performance drift, cybersecurity vulnerabilities, continuous learning algorithms. The framework handles them all the same way—through Intake and CCB classification.

16.1 The Pattern Holds

Whether the signal is a product feature, a bug, a complaint, a CVE, or an AI model retraining trigger, the pattern is identical:

1. Signal enters Intake
2. CCB classifies safety impact
3. Work proceeds with appropriate rigor
4. Evidence captured during sprint
5. Release Gate confirms completeness

The framework doesn't care about signal origin. It cares about safety impact and evidence.

16.2 AI/ML Considerations

For AI-enabled SaMD, additional considerations apply:

PCCP (Predetermined Change Control Plan): FDA guidance allows manufacturers to define, in advance, the types of changes that can occur without new submission. The PCCP defines boundaries.

Model retraining within PCCP boundaries: If your PCCP defines acceptable retraining parameters, changes within those boundaries follow a documentation-light path. The key is that the PCCP was pre-approved.

Model changes outside PCCP boundaries: Changes that exceed PCCP boundaries require RA engagement and potentially new submission. These are HIGH risk by definition.

The CCB classification question adapts: 'Does this model change stay within PCCP boundaries?' If yes, standard rigor (classification depends on other factors). If no, or if no PCCP exists, HIGH rigor plus RA consultation.

> **REGULATORY ANCHOR:** FDA's AI/ML-Based SaMD Action Plan and related guidance establish the PCCP framework. A well-defined PCCP enables iterative AI development within regulatory guardrails. Without a PCCP, every model change requires individual regulatory assessment.

16.3 Cybersecurity

When a CVE affects your system:

1. Security team identifies the vulnerability and severity
2. Signal enters Intake (urgently if critical)
3. CCB classifies—patient safety impact, not just security severity
4. If patient safety affected: HIGH classification regardless of CVE severity
5. If no patient safety impact: Classification based on other factors (may be LOW)
6. Expedited sprint or hotfix process with full rigor

The key distinction: CVE severity and safety impact are different assessments. A critical CVE in infrastructure with no patient-facing impact might be MEDIUM or LOW. A moderate CVE that could corrupt clinical data is HIGH.

> **REGULATORY ANCHOR**: FDA's 2023 Cybersecurity Guidance requires that cybersecurity risks be integrated with safety risk management. The framework accomplishes this by running CVEs through the same Risk Gate as any other change, with safety impact as the classification criterion. IEC 81001-5-1:2021 provides the cybersecurity lifecycle framework for health software, increasingly expected by EU Notified Bodies for MDR conformity assessment alongside FDA's guidance.

Causal discipline reminder: Environment contributes to hazardous situation; hazard remains the device characteristic (including data handling, logic, and availability design choices). A CVE or network compromise is the environmental condition that creates the hazardous situation — the hazard is always a device characteristic such as unencrypted storage, unvalidated input acceptance, or insufficient access controls.

16.4 Emerging Signal Types

As medical software evolves, new signal types will emerge:

Algorithm performance drift: AI systems may degrade over time as input data distributions shift. PMS monitoring should detect drift before it affects clinical outcomes.

Real-world evidence: Data from deployed devices may reveal patterns not visible in clinical trials. This evidence should feed back into risk estimates.

Interoperability issues: As devices connect to more systems, integration failures become a signal type requiring tracking.

The framework is signal-agnostic. As new signal types emerge, they enter Intake and flow through CCB. The classification criteria may evolve, but the process remains stable.

Health software scope expansion: IEC 82304-1:2016 establishes safety requirements for health software products — standalone software applications used in healthcare that may not meet the regulatory definition of a medical device. As IEC 62304 Edition 2 expands scope from 'medical device software' to 'health software' (see Appendix JJJ), the boundary between IEC 62304 and IEC 82304-1 becomes increasingly important. Organizations developing software that straddles the medical device/health software boundary should monitor this evolution and ensure their governance framework addresses both standards where applicable.

Chapter 16 Summary

AI, cybersecurity, and emerging signals all flow through the same process: Intake → CCB → Sprint → Release Gate.

For AI: PCCP boundaries determine whether changes stay within pre-approved parameters. Outside PCCP requires RA engagement.

For cybersecurity: CVE severity ≠ safety impact. Classification is based on patient safety, not just security severity.

The framework is designed to accommodate new signal types without structural change.

PART VI — INSPECTION DEFENSE

The real test of any compliance system is the inspection. Here's how to demonstrate confidence.

Chapter 17 — Facing the Inspector

Everything in this book leads to this moment: an FDA investigator or Notified Body auditor asks to see how you control design changes.

If you've implemented the framework, this conversation is straightforward. If you haven't, no amount of preparation will help.

17.1 The First Five Minutes

The first five minutes of any inspection set the tone. You have one chance to establish the frame.

Don't lead with documents. Lead with narrative.

Every change in our system—whether it's a feature, a bug fix, a complaint response, or a security patch—flows through our Risk Gate before entering a sprint. Our DQE classifies the safety impact, which determines the rigor level. During the sprint, evidence is captured as work happens. At release, we review evidence, not create it. Would you like me to walk you through a specific example?

That opening accomplishes several things:

- It demonstrates that you have a system
- It names specific roles and their accountabilities
- It addresses the temporal concern (during, not after)
- It offers to show evidence

17.2 The Example Walkthrough

When the inspector picks an example (or accepts your offer to choose one), walk them through the complete cycle:

1. Here's when the change entered Intake (Jira creation date)
2. Here's the CCB record showing DQE classification and rationale
3. Here's the Jira ticket with Safety_Impact, requirement links, risk links
4. Here's the risk file update dated during the sprint (not after)
5. Here's the test evidence with DQE approval timestamp
6. Here's the Sprint DHF packet showing all artifacts

Let the timestamps tell the story. They prove what you're claiming: evidence created during development, not reconstructed afterward.

17.3 Common Inspector Questions

MOCK INSPECTION Q&A: Who decided this was low risk?

> **Weak answer:** "Engineering thought it was low risk so we didn't do the full process."
>
> **Strong answer:** "DQE classified it at our CCB on March 3rd. The classification record shows the rationale: this change affects logging infrastructure only, with no user-facing or clinical impact. Here's the CCB record."

MOCK INSPECTION Q&A: Show me the risk analysis for this change.

> **Weak answer:** "Let me find the risk file... I think it was updated after the release."
>
> **Strong answer:** "The risk file was updated on March 8th during Sprint 47. Here's the specific hazard analysis entry. You can see it was updated three days before the code was finalized, and the hazard control matches the implementation."

MOCK INSPECTION Q&A: How do you ensure nothing ships without proper controls?

> **Weak answer:** "We have a process, and people follow it."
>
> **Strong answer:** "Our Jira workflow prevents items from being marked Done without DQE verification approval. Items cannot enter a sprint without CCB classification. Our Release Gate requires all items to show complete evidence before release proceeds. Here's the tracker showing current status—the workflow enforcement is built into the tool."

MOCK INSPECTION Q&A: What if someone bypasses the process?

Weak answer: "People wouldn't do that."

Strong answer: "The Jira workflow rules prevent bypass—items cannot move to In Progress without Safety_Impact. We also do monthly audits of completed items to verify evidence completeness. And frankly, the process is fast enough that bypassing it would take longer than following it."

KEY INSIGHT: Confidence in an inspection comes from knowing your system works—not from having documents that say it should work.

17.4 Notified Body Audit Considerations

For EU-marketed devices, Notified Body audits differ from FDA inspections in structure and emphasis. While FDA inspectors focus on facility-based, process-flow assessment ('show me how a change moved through your system'), Notified Bodies conduct detailed Technical Documentation reviews (Annex II/III), assess General Safety and Performance Requirements (GSPR) compliance against Annex I, and evaluate PMS and PMCF plans.

The framework prepares for both: Sprint DHF Packets compile directly into Technical Documentation Annex II sections (Section 3 for design and manufacturing information, Section 4 for GSPR compliance evidence, Section 6 for verification and validation). The risk file's hazard-to-GSPR traceability demonstrates Annex I compliance. The PMS loop addresses Articles 83-86 documentation requirements.

When facing a Notified Body audit, lead with the same narrative approach described in Section 17.1 but emphasize the Technical Documentation structure and GSPR traceability rather than the process flow. Show how sprint evidence compiles into the structured documentation the Notified Body reviews. See Appendix LLL.3 for detailed Notified Body audit alignment.

ROLE ACCOUNTABILITY: For Notified Body audits: RA leads the audit interface and owns Technical Documentation submission. DQE provides design control evidence and risk management file access. PM coordinates Sprint DHF Packet compilation into Annex II structure. PMS Manager presents PMS Plan and PSUR data. Engineering Lead addresses technical architecture questions.

17.5 When You Don't Know the Answer

Sometimes an inspector will ask a question you can't answer immediately. How you respond matters.

Good response: 'I don't have that information immediately available. Let me find the right person and get back to you within the hour.'

Bad response: 'I'm not sure, but I think...' followed by speculation.

Inspectors prefer honest uncertainty to confident incorrectness. If you don't know, say so and commit to finding out.

> **483 PATTERN:** Inconsistent answers from different team members about the same process. Inspector asks the same question to three people and gets three different answers. The process is either not defined or not followed.

Chapter 17 Summary

Lead with narrative, not documents. Establish that you have a system before showing artifacts.

Let timestamps tell the story. Evidence dated during development is far more credible than evidence dated at release.

Answer questions directly with specific evidence. Reference roles, dates, and records—not general assertions about following process.

When you don't know, say so and commit to finding out. Never speculate.

For Notified Body audits, emphasize Technical Documentation structure and GSPR traceability. Sprint DHF Packets compile directly into Annex II sections. RA leads the audit interface.

Chapter 18 — Common Failure Modes

Even organizations that implement this framework can fall into patterns that undermine it. Here are the most common failure modes and how to avoid them.

18.1 CCB Rubber-Stamping

Symptom: Every item is classified LOW. CCB takes 5 minutes for 30 items.

Root cause: DQE isn't actually analyzing items. Classification has become a checkbox exercise.

Detection: Audit a sample of classifications. Review items classified LOW and verify the classification was appropriate. Look for items that affect user-facing functionality but were classified LOW.

Fix: Retrain DQE on the universal question. Review classification criteria. Consider requiring brief written rationale for LOW classifications. **Consequence if uncorrected:** If audit reveals systematic misclassification, the organization must revert to Full Oversight Mode—DQE manually reviews every item before sprint entry until CCB demonstrates consistent, defensible classification over at least two sprint cycles. This is not punitive; it's protective. Document the audit finding, the remediation period, and the return-to-normal criteria.

18.2 Engineering-Only Exceptions

Symptom: Certain changes bypass CCB because they're 'just engineering.'

Root cause: Cultural resistance to classifying refactoring, performance work, or infrastructure changes.

Detection: Review items that entered sprints without Safety_Impact field populated. Review infrastructure or refactoring work for classification.

Fix: Enforce Jira workflow rules that prevent sprint entry without classification. Educate teams that 'engineering-only' doesn't exist—even infrastructure changes can affect device behavior.

Governance consequence: Engineering-only bypass is a quality system nonconformance requiring documented investigation and systemic correction — not an informal coaching moment. When bypass is discovered, the affected change must be quarantined from release until properly classified by DQE, the bypass event is documented as a CAPA input, and root cause is addressed. Inspectors who find bypass patterns will ask about your self-monitoring capability — the CAPA record is your answer.

18.3 Evidence Reconstruction

Symptom: All documentation timestamps cluster at sprint end or release time.

Root cause: Teams are waiting until work is done to create evidence. DQE isn't embedded in sprints.

Detection: Plot timestamp distributions for SRS updates, risk file updates, and verification approvals. Look for clustering patterns.

Fix: Embed DQE in sprint standups. Require incremental evidence during sprints, not just at the end. Review timestamp distributions in retrospectives.

18.4 Tracker Gaming

Symptom: Tracker shows all GREEN, but actual evidence is incomplete.

Root cause: Status fields updated without verifying underlying evidence exists.

Detection: Spot-check. Pick a random GREEN item and verify the linked evidence actually exists and is complete.

Fix: Enforce that status can only change when evidence links are populated. Daily verification by PM. Random audits by quality.

18.5 CCB Bottleneck

Symptom: CCB queue grows faster than it's processed. Items wait weeks for classification.

Root cause: CCB meets too infrequently or processes items too slowly.

Detection: Track CCB queue age. If items routinely wait more than a few days, there's a capacity problem.

Fix: Increase CCB frequency. Streamline classification criteria. Pre-triage items before CCB. Delegate clear-cut classifications to DQE alone (with CCB ratification).

18.6 Role Confusion

Symptom: Multiple people think they're accountable for the same decision. Or no one is.

Root cause: Role definitions aren't clear or aren't followed.

Detection: Ask three people 'who decides risk classification?' If you get three different answers, you have role confusion.

Fix: Document and communicate role definitions. Post the RACI matrix. Enforce role boundaries—people outside their role don't make decisions for that role.

Chapter 18 Summary

Failure modes are predictable: rubber-stamping, exceptions, reconstruction, gaming, bottlenecks, and role confusion.

Detection requires active monitoring: audit samples, review timestamps, spot-check evidence, track queue age, verify role clarity.

Fixes address root causes, not symptoms. Workflow enforcement, DQE embedding, and clear role definitions prevent most failure modes.

Governance Escalation: When failure modes are detected through internal audit, the response must be proportionate to the risk. Minor issues warrant coaching. Systematic failures warrant temporary reversion to higher oversight until the root cause is resolved and verified. Document all findings and corrective actions—inspectors will ask about your self-monitoring capability.

PART VII — THE FUTURE

The organizations that master this approach won't just be compliant—they'll be faster, more reliable, and better positioned for whatever regulatory evolution comes next.

Chapter 19 — Where This Goes From Here

The framework in this book isn't static. It will evolve as the industry evolves, as regulations evolve, as technology evolves.

19.1 Continuous Learning Systems

AI-enabled devices that learn from field data will accelerate the loop. Signals won't wait for complaint reports—they'll be detected automatically. Performance drift will trigger alerts. The CCB will need to operate faster, perhaps continuously.

The framework supports this evolution because it separates classification (a decision) from meetings (an implementation choice). If classification can happen in real-time based on clear criteria, the CCB meeting becomes a review mechanism, not a decision point. Automated classification within pre-defined criteria, with CCB oversight for edge cases.

19.2 Regulatory Evolution

Regulations will continue to evolve. EU MDR brings different requirements than FDA. New guidance on AI/ML will create new expectations. International harmonization will simplify some things and complicate others.

The framework survives regulatory evolution because it's built on principles that regulations share: documented design inputs, systematic risk management, verified outputs, traceable development. The specific requirements change; the principles don't.

19.3 Competitive Advantage

Organizations that master native design control will have a significant competitive advantage. They'll release faster because there's no documentation catch-up. They'll audit better because evidence already exists. They'll adapt faster because their systems are designed for change.

The organizations that treat compliance as overhead will find themselves unable to compete. Not because they're less capable—but because they've chosen to carry the retrofit tax while others haven't.

19.4 The Ultimate Goal

The ultimate goal is to make compliance invisible. Not by removing rigor, but by embedding it so deeply in how work gets done that no one thinks about it separately.

When an engineer makes a design decision, the risk analysis update happens naturally. When a test passes, the evidence is captured automatically. When a sprint ends, the design history is complete without anyone assembling it.

That's when you know the framework has succeeded: when no one mentions compliance anymore because it's just part of how work gets done.

> **KEY INSIGHT:** The goal isn't to be great at compliance. The goal is to be great at building safe, effective medical software—and let compliance be a natural byproduct.

Chapter 19 Summary

The framework evolves with the industry. Continuous learning systems will accelerate the loop. Regulatory evolution will bring new requirements but the principles remain stable.

Organizations that master native design control gain competitive advantage: faster releases, better audits, greater adaptability.

The ultimate goal is invisible compliance — embedded so deeply in how work gets done that no one thinks about it separately.

Epilogue — A New Mental Model

If you've read this far, you've absorbed a lot of information. Let me leave you with the simplest possible summary.

The old mental model:

> *We build software. Then we prove it's compliant.*

The new mental model:

> *We build software compliantly. Evidence is a byproduct of how we work.*

That's it. That's the entire framework in two sentences.

Everything else—the CCB, the tracker, the roles, the terminology—exists to make that mental model practical. To create an environment where building software and creating evidence are the same activity, not sequential phases.

The framework isn't easy to implement. It requires changes to tools, processes, and culture. It requires rethinking assumptions that many organizations have held for years. It requires people to accept new accountabilities and give up old habits.

But once implemented, the framework becomes invisible. Teams stop thinking about compliance and start thinking about building great software. The compliance happens automatically because it's built into how work gets done.

That's when you know you've succeeded: when no one mentions compliance anymore because it's just part of how work gets done. When the inspector asks about your system and you can walk through any change end-to-end with confidence. When releases ship on schedule because there's no documentation catch-up. When engineering and quality are partners, not adversaries.

That state is achievable—whether you're building for FDA clearance, EU MDR conformity, or both. This book shows you how.

Good luck.

APPENDICES

Appendix Finder

Use this guide to find the appendix you need by what you're trying to do.

Getting Started & Quick Reference

A — 90-Day Implementation Quick Start Guide
K — Engineering Leader Quick Reference
L — Chapter Summaries for Quick Reference
E — Glossary of Terms
II — Frequently Asked Questions

Walkthroughs & Worked Examples

B — Complete Walkthrough: Bug from Discovery to Release
H — Complete Walkthrough: HIGH Risk Feature
J — Complete Walkthrough: Complaint to Resolution
SS — Case Study: Framework Implementation
HHH — Book Example Tracker Views
YY — Usability Engineering Process Walkthrough

Templates, Tools & Trackers

GGG — Comprehensive Design Control Tracker
M — Sprint DHF Packet Template
HH — Document Templates
BBB — Software Development Plan Template
FF — Detailed Jira Configuration Guide
III — Essential SOP Architecture
MMM — Design Control Deliverables by Phase (Risk-Scaled)
NNN — Consolidated Governance Template Library

Inspection Preparation & 483 Defense

F — Comprehensive 483 Pattern Library
G — Extended Mock Inspection Q&A Library
GG — Extended Mock Inspection Scenarios
OO — Inspection Preparation Deep Dive
RR — Extended 483 Pattern Analysis
X — Audit Preparation Checklist

Risk Management & Classification

P — Risk File Structure and Maintenance

T — Detailed CCB Classification Scenarios
EEE — Quantitative P1/P2 Probability Estimation
FFF — Use-Related Risk Analysis: Connecting HF and Risk Management

Regulatory Mapping & Compliance

D — Regulatory Reference Summary
TT — FDA QMSR and ISO 13485:2016 Integration
UU — ISO 13485:2016 Clause-by-Clause Framework Mapping
VV — Updated Regulatory Reference: QMSR
I — EU MDR Compliance Mapping
LLL — EU MDR/IVDR Comprehensive Framework Alignment
AAA — IEC 62304 Software Lifecycle Integration
JJJ — IEC 62304 Edition 2 Transition Advisory

Usability & Human Factors Engineering

WW — IEC 62366-1 Usability Engineering Integration
XX — AAMI HE75:2025 Human Factors Engineering
KKK — AAMI HE75:2025 Updated Guidance
ZZ — Usability Engineering Regulatory References

Engineering Practices & Process

C — RACI Matrix
CCC — Operational Role Clarity: De-Coupling QA/RA
DDD — PM Workflow Enforcement Rules
N — Tool Validation Guidance
O — SRS Authoring for Regulated Software
Q — Verification Testing Best Practices
R — Validation vs. Verification: When and How
PP — Statistical Methods in Design Verification
QQ — Design Transfer for SaMD
U — Release Management Scenarios
V — Design Review Implementation
Y — Sprint Ceremonies in Regulated Environment
Z — Metrics and Measurement
BB — CAPA Integration with Design Control
NN — Documentation Management Best Practices

Specialized Topics

JJ — Clinical Decision Support (CDS) Considerations
KK — AI/ML Software as Medical Device
S — Cybersecurity Risk Integration

LL — Interoperability and Integration
MM — Mobile and Distributed Software
CC — Supplier and Third-Party Component Control
DD — Integrating Legacy Systems

Organizational Scaling & Change

AA — Organizational Change Management
EE — Scaling for Different Organization Sizes
W — Training and Competency Requirements

Large-Format Reference

OOO — Large-Format Figure and Table Reference (rotate device 90° clockwise to view)

Appendix A — Quick Start Guide: 90-Day Implementation

If you want to implement this framework, here's a practical 90-day roadmap with specific actions for each phase.

Week 1: Foundation

□ Identify your DQE (or assign someone to that accountability)

□ Create the 5 custom fields in Jira: Safety_Impact, CCB_Reference, SRS_Link, Risk_Link, Verification_Status

□ Configure Safety_Impact as required field for sprint entry

□ Schedule your first CCB meeting—weekly recurring

□ Invite: DQE, PM, Engineering Lead, Product Owner

□ Adopt the Terminology Lock from Chapter 6—print and post

Week 2-4: Pilot Sprint

□ Run ONE sprint with the full model

□ Classify every item at CCB before sprint entry

□ DQE attends standups (observes, speaks only when needed)

□ Apply Definition of Done with three sign-offs (Engineering, Product, DQE)

□ Generate a Sprint DHF Packet at sprint close

□ Retrospective: What worked? What was friction? What needs adjustment?

Week 5-8: Refinement

□ Adjust CCB cadence based on volume (twice-weekly if needed)

□ Create Jira workflow rules to enforce classification before sprint entry

□ Train Product Owners on SRS linkage expectations

□ Train Engineering on what 'verification evidence' means

□ Build your first tracker dashboard (even in Excel)

□ Document lessons learned from pilot sprints

Week 9-12: Scale

□ Expand to additional teams

□ Integrate PMS signals into Intake (complaints enter same process)

□ Formalize Release Gate checklist

□ Conduct mock inspection: 'Show me how this change was controlled'

□ Document organizational procedures

□ Train additional DQEs if needed for scale

Success Criteria at 90 Days

✓ Every sprint item has Safety_Impact before entering sprint

✓ DQE approval timestamp exists for all MEDIUM/HIGH items

✓ Sprint DHF Packet generated with minimal manual effort

✓ Release Gate is confirmatory (no surprises)

✓ Team can answer: 'Show me the sprint that implemented this change'

✓ Timestamp distributions show parallel execution, not clustering

Common Week 1 Mistakes to Avoid

✗ Don't try to classify the entire backlog at once—just next sprint items

✗ Don't create a separate 'compliance system'—extend Jira

✗ Don't skip CCB for 'small' items—classify everything

✗ Don't make DQE a gatekeeper—embed them in the flow

✗ Don't expect perfection—expect learning

Appendix B — Walkthrough: A Bug from Discovery to Release

This appendix provides a complete day-by-day walkthrough of how a bug flows through the framework.

Figure B-1: BUG-4521 Complete Lifecycle — From Discovery to Resolution

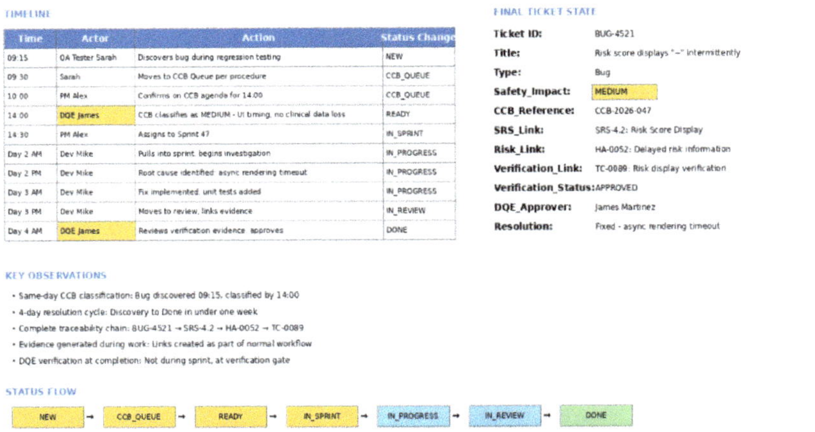

Figure B-1: BUG-4521 Complete Lifecycle — From Discovery to Resolution

Scenario

QA discovers that the risk score display occasionally shows '--' instead of a numeric value when the calculation takes longer than expected. This has been seen 3 times in testing.

Day 0 (Monday): Discovery & Intake

09:15 — QA Engineer Sarah discovers the bug during regression testing. She creates Jira ticket BUG-4521: 'Risk score displays -- intermittently during slow calculations'

09:30 — Sarah moves item to CCB Queue status per procedure. Bug cannot be worked until classified.

10:00 — PM Alex sees the CCB queue item and confirms it's on the afternoon CCB agenda.

Day 0 (Monday): Risk Gate / CCB

14:00 — CCB convenes. Attendees: DQE Maria (classification authority), PM Alex (organizer), Engineering Lead Tom, Product Owner Lisa.

14:02 — PM Alex presents BUG-4521: 'Display issue where score shows dashes intermittently.'

14:03 — DQE Maria asks the universal question: 'Does this affect clinical workflow or patient safety?'

14:05 — Discussion: The '--' display could cause clinician to miss a risk score. Clinician might delay decision waiting for score, or proceed without score, missing a risk signal.

14:08 — DQE Maria states classification: 'MEDIUM RISK. Not HIGH because existing data is not corrupted and there's a visible indicator that something is wrong. Not LOW because clinical display is affected. Rationale documented.'

▸ CCB CLASSIFICATION RATIONALE (causal chain form — required for MEDIUM)

Hazard: Risk score display rendering — software permits '--' output under timeout/load conditions without graceful recovery.

Hazardous Situation: Clinician views '--' instead of a numeric risk score; exposed to missing clinical information at a decision point.

P1 basis: Observed at 3/500 test runs (0.6%); estimated higher under production load.

Harm: Delayed clinical decision based on incomplete information.

Most likely harm: workflow delay, manual retry.

Worst plausible harm: missed deterioration signal in time-critical context.

P2 basis: Existing recovery pathway (manual refresh) reduces harm probability.

Clinician can request retry; data is not lost. Estimated P2: low to moderate.

Safety_Impact = MEDIUM. Risk file entry: HA-0023 (Incomplete Display).

14:10 — PM Alex records: Safety_Impact = MEDIUM, CCB_Reference = CCB-2024-0892

14:12 — BUG-4521 moves from CCB Queue to Backlog. Ready for sprint planning.

Day 1 (Tuesday): Sprint Planning

09:00 — Sprint 47 planning begins. BUG-4521 is in the prioritized backlog.

09:20 — Product Owner Lisa confirms priority based on user impact and CCB classification.

09:25 — Engineering Lead Tom estimates: 3 story points. Root cause investigation + fix + testing.

09:30 — DQE Maria confirms: 'MEDIUM RISK means I need to review the verification approach and approve test results before Done.'

09:35 — BUG-4521 added to Sprint 47 commitment. Required fields confirmed: Safety_Impact = MEDIUM, CCB_Reference populated.

Days 2-3 (Wed-Thu): Parallel Execution

ENGINEERING (Tom):

• Wednesday AM: Root cause identified—timeout not handled gracefully

• Wednesday PM: Fix approach defined: Add loading state, extend timeout, add retry

• Thursday: Implementation complete, unit tests passing, code review submitted

PRODUCT (Lisa):

• Wednesday: Confirms SRS section 4.2.3 (Display Requirements) applies

• Wednesday: No SRS change needed—existing requirement covers expected behavior, this is a defect against that requirement

• Thursday: Adds SRS_Link to Jira ticket

DQE (Maria):

• Wednesday: Reviews hazard analysis—existing HA-0023 (Incomplete Display) covers this scenario

• Wednesday: Updates risk file: P1 estimate refined based on actual occurrence rate (3 in 500 test runs = 0.6%)

• Thursday: Confirms existing risk control (display validation) applies. Adds Risk_Link to Jira ticket

Day 4 (Friday): Verification

09:00 — Tom executes targeted verification tests:

• TC-4521-01: Verify loading state displays during calculation — PASS

• TC-4521-02: Verify retry logic after timeout — PASS

• TC-4521-03: Verify final score displays correctly — PASS

• TC-4521-04: Reproduce original bug scenario (forced delay) — PASS (bug no longer occurs)

11:00 — Tom adds test evidence links to BUG-4521 in Jira

14:00 — DQE Maria reviews test evidence:

• Confirms test cases cover the risk control verification

• Confirms results demonstrate fix effectiveness

• Sets Verification_Status = APPROVED with timestamp

Day 4 (Friday): Definition of Done

15:00 — Sprint standup reviews BUG-4521 for Done status:

✓ Engineering (Tom): Code complete, reviewed, unit tests passing, integration tests passing

✓ Product (Lisa): SRS linkage confirmed, no SRS update needed

✓ DQE (Maria): Risk file updated, verification approved, traceability complete

15:10 — All three sign-offs confirmed. BUG-4521 status → DONE

The Evidence Trail

The Sprint 47 DHF Packet contains for BUG-4521:

- CCB record CCB-2024-0892 with DQE classification and rationale (dated Monday)
- Jira ticket with all required fields populated
- Risk file update showing P1 refinement and control verification (dated Wednesday-Thursday)
- Test execution results with PASS status (dated Friday)
- DQE approval signature and timestamp (dated Friday)
- Traceability: BUG-4521 → SRS 4.2.3 → HA-0023 → TC-4521-*

Timestamp Analysis

An inspector reviewing this change would see:

Monday 14:10 — CCB classification (before sprint)

Tuesday 09:35 — Sprint planning commitment

Wednesday — Implementation begins, SRS link added, Risk file update begins

Thursday — Implementation complete, Risk file complete

Friday 11:00 — Test execution complete

Friday 14:00 — DQE approval

Friday 15:10 — Definition of Done confirmed

Activity distributed across sprint. Classification BEFORE implementation. Evidence DURING work. This is native design control.

Appendix C — RACI Matrix

This matrix defines accountability for key framework activities.

Table C-1: RACI Matrix — Framework Activities

Activity	PO	Eng	DQE	RA	PM	PgM	PMS
Feature definition	A	C	I	I	I	I	I
Bug triage	C	A	I	I	R	I	I
CCB classification	I	C	A	C	R	I	I
Requirements (SRS)	A	C	R	I	I	I	I
Design (SDS)	C	A	I	I	I	I	I
Risk file maintenance	I	C	A	C	I	I	C
Verification approval	I	R	A	I	I	I	I
Release decision	I	I	R	R	R	A	I
CCB facilitation	C	C	C	C	A	I	I
Tracker maintenance	R	R	R	I	A	I	I
Regulatory submission	I	C	C	A	I	R	I
Complaint triage	I	I	C	I	I	I	A
CAPA initiation	I	I	C	I	I	C	A
Field signal analysis	I	I	C	I	I	I	A
SOUP register maintenance	I	R	A	I	I	I	I
IEC 62304 safety class determination	I	C	A	C	I	I	I
Usability Engineering File (UEF)	C	C	A	I	I	I	I

Activity	PO	Eng	DQE	RA	PM	PgM	PMS
Formative usability evaluation	C	C	A	I	I	I	I
Summative usability evaluation	C	R	A	C	R	C	I
EU Technical Documentation	C	C	R	A	R	I	I
GSPR compliance mapping	I	C	R	A	I	I	I
EU MDR PMS Plan/PSUR	I	I	C	C	I	I	A
Notified Body audit support	I	C	R	A	R	I	C
PCCP boundary monitoring (AI/ML)	C	C	R	A	I	I	C
Cybersecurity risk assessment	I	R	A	C	I	I	I
Legacy software gap analysis	I	R	A	C	I	I	I

A **Accountable** — *Owns the outcome (single person)*
R **Responsible** — *Does the work*
C **Consulted** — *Provides input before decision*
I **Informed** — *Kept aware after decision*

Appendix D — Regulatory Reference Summary

21 CFR 820.30 — Design Controls

820.30(a): General — establishes the scope of design controls and the device classes subject to them

820.30(b): Design and development planning — requires plans that describe activities and assign responsibilities

820.30(c): Design input — requires documented design input requirements

820.30(d): Design output — requires documented design outputs that meet input requirements

820.30(e): Design review — requires planned and documented reviews

820.30(f): Design verification — requires confirmation that outputs meet inputs

820.30(g): Design validation — requires confirmation that device meets user needs

820.30(h): Design transfer — requires procedures for transfer to production

820.30(i): Design changes — requires procedures for identifying, documenting, validating, and approving changes

820.30(j): Design history file — requires a file containing or referencing records demonstrating compliance

ISO 14971:2019 — Risk Management

Clause 4.1: Risk management process — ongoing throughout lifecycle

Clause 5: Risk analysis — identification of hazards and hazardous situations

Clause 6: Risk evaluation — determination of acceptability

Clause 7: Risk control — selection and implementation of measures

Clause 8: Overall residual risk evaluation

Clause 9: Risk management review

Clause 10: Production and post-production activities

IEC 62304:2006+A1:2015 (Edition 1.1) — Software Lifecycle

Clause 5.1: Software development planning

Clause 5.2: Software requirements analysis

Clause 5.3: Software architectural design

Clause 5.4: Software detailed design

Clause 5.5: Software unit implementation and verification

Clause 5.6: Software integration and integration testing

Clause 5.7: Software system testing

Clause 5.8: Software release

Clause 6: Software maintenance

Clause 9: Software problem resolution

Clause 4.3: Software safety classification — assignment of Class A, B, or C based on severity of potential harm

Clause 4.4: Legacy software — compliance pathway for software not developed under IEC 62304

Clause 7: Software risk management — integration with ISO 14971 for software-specific hazards

Clause 8: Software configuration management — identification and control of configuration items including SOUP

AAMI TIR45:2023 — Agile Guidance

TIR45 establishes that iterative development methodologies are compatible with IEC 62304. Key guidance includes:

• Requirements can evolve during development

• Risk analysis can be updated iteratively

• Verification can happen throughout development

• Design history can be assembled from iterative artifacts

IEC 62366-1:2015/AMD1:2020 — Usability Engineering

Clause 5.1: Use specification — identifying intended users, use environments, and user interface

Clause 5.7: User interface evaluation plan — planning formative and summative evaluation

Clause 5.8: Formative evaluation — iterative usability assessment during design

Clause 5.9: Summative evaluation — formal evaluation on release candidate with representative users

Clause 5.10: User interface of unknown provenance (UOUP) — alternative evaluation per Annex C

ISO 13485:2016 — Quality Management System

Clause 7.3.1: Design and development planning — planning output updated as development progresses

Clause 7.3.2: Design and development inputs — requirements relating to device function, safety, and regulatory

Clause 7.3.3: Design and development outputs — outputs meeting inputs, providing essential information

Clause 7.3.4: Design and development review — systematic review at suitable stages

Clause 7.3.5: Design and development verification — confirmation outputs meet inputs

Clause 7.3.6: Design and development validation — confirmation device meets user needs and intended use

Clause 7.3.7: Design and development transfer — procedures for transfer to production

Clause 7.3.8: Design and development changes — identification, review, verification, validation, approval

Clause 7.3.9: Design and development files — DHF containing or referencing records

Appendix E — Glossary of Terms

CCB: Change Control Board — the governance mechanism that classifies changes by safety impact before sprint entry

Design Control: The systematic process required by 21 CFR 820.30 to control design from concept through transfer

Design History File (DHF): The compilation of records describing the design history of a finished device

DQE: Design Quality Engineer — role accountable for risk classification and verification adequacy at sprint cadence

GSPR: General Safety and Performance Requirements — EU MDR/IVDR Annex I requirements that medical devices must meet to demonstrate safety and performance

Harm: Physical injury or damage to health, property, or environment

Hazard: Potential source of harm; always a device characteristic, never a user action

Hazardous Situation: Circumstance where people, property, or environment are exposed to a hazard

HIGH Risk: Classification for changes affecting clinical decisions or creating new hazardous situations

Intake: Single entry point for all change signals regardless of source

Legacy Software: Software already developed and in use that was not developed under IEC 62304. Clause 4.4 provides a compliance pathway requiring gap analysis, risk assessment, and remediation planning

LOW Risk: Classification for changes with no clinical or user-facing impact. LOW does not mean 'no controls'—verification is still required, and the classification decision itself must be documented.

MEDIUM Risk: Classification for changes affecting user workflow but not clinical decisions directly

Native Design Control: The practice of generating design control evidence during development rather than retrofitting it afterward. Evidence emerges as a natural byproduct of engineering workflow, not as a separate compliance activity

P1: Probability that hazardous situation occurs (exposure probability)

P2: Probability that hazardous situation leads to harm (given exposure)

Parallel Execution: Practice of evolving requirements, risk analysis, and verification simultaneously

PCCP: Predetermined Change Control Plan — FDA mechanism for pre-approving types of AI/ML changes

PMS: Post-Market Surveillance — systematic monitoring of device performance after market release

PSUR: Periodic Safety Update Report — EU MDR Article 86 requirement for Class IIa/IIb/III devices summarizing PMS data, benefit-risk conclusions, and PMCF findings

RA: Regulatory Affairs — role accountable for regulatory strategy and submission decisions

Release Gate: Checkpoint confirming evidence completeness before release

Regulated Sprint: A sprint in which design control activities (classification, risk analysis, verification, documentation) execute in parallel with development rather than sequentially after it. The fundamental unit of work in this framework

Retrofit Tax: Hidden cost of doing compliance after development instead of during

Risk Gate: CCB mechanism classifying safety impact before work enters sprint

Safety_Impact (HIGH / MEDIUM / LOW): A governance classification set by the CCB that determines the rigor of design controls applied to a change. Aligned to risk-based thinking but not the full ISO 14971 risk estimate. Determines what evidence a change requires

Software Safety Class: IEC 62304 classification (A, B, or C) based on severity of potential harm if software contributes to a hazardous situation. Class C = death/serious injury, Class B = non-serious injury, Class A = no contribution to hazardous situation

SOUP: Software of Unknown Provenance — software already developed and generally available that is incorporated into a medical device but was not developed under IEC 62304. Includes open-source libraries, commercial SDKs, and off-the-shelf components

Sprint DHF Packet: Evidence package produced at sprint close containing all design history for that sprint

SRS: Software Requirements Specification — documented requirements for the software

Tracker: Design Control Tracker providing real-time visibility into compliance status

UEF: Usability Engineering File — compilation of usability engineering records per IEC 62366-1, maintained parallel to the DHF and cross-referenced to the risk management file

Universal Question: CCB classification question: 'Does this affect clinical workflow or patient safety?'

Validation: Confirmation that device meets user needs — 'Did we build the right thing?'

Verification: Confirmation that design outputs meet design inputs — 'Did we build it right?'

Appendix F — Comprehensive 483 Pattern Library

This appendix provides an extended library of 483 observation patterns commonly encountered in SaMD inspections. Each pattern includes the observation, root cause, and prevention strategy.

F.1 Design Input Patterns

483 PATTERN: Requirements documents that don't match software behavior. Inspectors find discrepancies between SRS and actual implementation. Root cause: Requirements updated after implementation, or implementation diverged without updating requirements. Prevention: Parallel execution with Product Owner verifying SRS accuracy during sprint.

483 PATTERN: Missing or incomplete intended use definition. Device descriptions lack clear clinical context. Root cause: Product team focused on features, not clinical application. Prevention: Require intended use statement in SRS, reviewed by RA before CCB approval.

483 PATTERN: User needs not traceable to requirements. Design inputs exist, but cannot trace back to user needs or market analysis. Root cause: Requirements written without formal user needs elicitation. Prevention: Require user needs reference for each SRS section.

F.2 Risk Management Patterns

483 PATTERN: Risk analysis performed after development decisions. Inspector finds risk file dated weeks after code commits. Root cause: DQE not embedded in sprints, risk analysis treated as documentation rather than design input. Prevention: Require risk file updates during sprint, not after.

483 PATTERN: Hazards listed as 'user error' without device characterization. Risk file shows user actions as hazards. Root cause: Misunderstanding of ISO 14971 causal chain. Prevention: Terminology lock training, DQE review of hazard entries.

483 PATTERN: Risk controls not verified. Risk file lists controls, but no evidence they were tested. Root cause: Disconnect between risk management and verification. Prevention: Traceability from risk controls to test cases, DQE verification approval.

483 PATTERN: P1/P2 estimates not based on data. Probability estimates appear arbitrary. Root cause: Risk analysis performed without operational data or testing results. Prevention: Require data justification for probability estimates, update with usability and field data.

483 PATTERN: Residual risk not explicitly evaluated. Risk file shows controls implemented but no residual risk assessment. Root cause: Skipped Clause 8 overall residual risk evaluation. Prevention: Require residual risk evaluation at each release gate.

F.3 Change Control Patterns

483 PATTERN: Changes implemented without CCB review. Items entered sprints without classification. Root cause: CCB bypassed for 'urgent' items or 'engineering-only' changes. Prevention: Jira workflow enforcement, no sprint entry without Safety_Impact.

483 PATTERN: CCB records lack classification rationale. Records show decisions but not reasoning. Root cause: CCB treated as rubber-stamp approval rather than risk classification. Prevention: Require written rationale for every classification.

483 PATTERN: Inconsistent classification criteria. Similar changes classified differently without explanation. Root cause: No calibration process for DQE decisions. Prevention: Classification reference library, periodic calibration reviews.

483 PATTERN: 'Engineering-only' changes found to affect device behavior. Items claimed to have no user impact actually affected clinical functionality. Root cause: Classification based on requester assertion, not analysis. Prevention: DQE independence, universal question applied consistently.

F.4 Verification Patterns

483 PATTERN: Test cases not traceable to requirements. Tests exist but cannot be linked to specific requirements. Root cause: Testing developed independently of requirements. Prevention: Require SRS_Link for all test items, traceability validation before release.

483 PATTERN: Regression testing not documented after changes. New features tested but unchanged functionality not re-verified. Root cause: Regression testing informal or undocumented. Prevention: Defined regression strategy, documented execution for each release.

483 PATTERN: Test evidence does not demonstrate requirement fulfillment. Test passed but unclear how it verifies the requirement. Root cause: Test cases too generic or acceptance criteria vague. Prevention: Test cases must reference specific acceptance criteria and expected results.

483 PATTERN: Unit test coverage claims not verified. Organization claims high coverage but cannot demonstrate. Root cause: Coverage metrics not captured or retained. Prevention: Automated coverage reporting, coverage thresholds in Definition of Done.

F.5 Design Review Patterns

483 PATTERN: Design reviews conducted without quality participation. Engineering reviews exist but no cross-functional involvement. Root cause: Code review conflated with design review. Prevention: Require DQE attendance at design reviews, documented outcomes.

483 PATTERN: Design review findings not tracked to closure. Issues identified but no evidence of resolution. Root cause: Reviews conducted but no follow-up mechanism. Prevention: Design review actions tracked in Jira, closure required before release.

483 PATTERN: No evidence of design review at required stages. Process requires reviews but no records exist. Root cause: Reviews scheduled but not conducted, or conducted informally. Prevention: Design review as Definition of Done checkpoint.

F.6 DHF and Records Patterns

483 PATTERN: DHF assembled retroactively before inspection. Evidence of DHF creation dates near audit rather than during development. Root cause: DHF treated as audit preparation rather than ongoing compilation. Prevention: Sprint DHF packets generated at sprint close.

483 PATTERN: Records conflict between Jira and DHF documents. Different information in engineering system vs. compliance documents. Root cause: DHF maintained separately from development records. Prevention: DHF generated from Jira, single source of truth.

483 PATTERN: Missing approval signatures on critical documents. Documents exist but approval not documented. Root cause: Informal approval processes, assumption that completion implies approval. Prevention: Workflow-enforced approval, electronic signatures with timestamps.

Appendix G — Extended Mock Inspection Q&A Library

This appendix provides additional mock inspection questions and answer guidance for common regulatory inquiries.

G.1 Design Input Questions

MOCK INSPECTION Q&A: How do you ensure requirements are complete before implementation?

> **Weak answer:** "We review them in sprint planning."
>
> **Strong answer:** "We don't require requirements to be 'complete' before implementation—they evolve during the sprint. Our CCB classifies items before sprint entry, which ensures safety considerations are documented. During the sprint, Product Owner refines requirements as engineering reveals implications. By sprint end, requirements are complete and verified against implementation."

MOCK INSPECTION Q&A: Show me how user needs translate to software requirements.

> **Weak answer:** "Our product team defines requirements based on customer feedback."
>
> **Strong answer:** "Here's our traceability from user needs to SRS. This user need document captures clinical workflow requirements from our formative research. Each user need links to one or more SRS requirements. Here's an example: User Need UN-015 'Clinician needs to view risk score during triage' links to SRS sections 4.2.1 through 4.2.4 covering display, format, timing, and alerts."

G.2 Risk Management Questions

MOCK INSPECTION Q&A: How do you identify hazards for software?

Weak answer: "We brainstorm potential issues during development."

Strong answer: "We use a systematic approach based on ISO 14971. For each clinical function, we identify device characteristics that could contribute to harm. We ask: what could this software do wrong that could affect patient care? Our hazard categories include: wrong patient, wrong result, wrong timing, missing information, and unintended action. Here's our hazard identification checklist and how it was applied to this feature."

MOCK INSPECTION Q&A: How do you determine if a risk is acceptable?

Weak answer: "We discuss it and make a judgment."

Strong answer: "We use defined acceptance criteria based on our risk matrix. Here's the matrix showing severity categories and probability levels. The acceptance boundary is pre-defined—risks above the line require additional controls or design changes. Residual risks in the yellow zone require documented benefit-risk analysis. This specific risk was evaluated against these criteria—here's the evaluation record showing the P1 estimate, P2 estimate, severity, and acceptability determination."

MOCK INSPECTION Q&A: How do you verify risk controls are effective?

Weak answer: "We test the software."

Strong answer: "Risk controls are verified through targeted testing. Each risk control in the risk file links to specific test cases designed to verify that control. For example, Risk Control RC-023 'Display timeout warning before data refresh' links to test cases TC-4521-01 through TC-4521-03 which specifically verify the warning behavior under various conditions. Here's the traceability and here are the test results showing the control is effective."

G.3 Change Control Questions

MOCK INSPECTION Q&A: How do you evaluate changes for safety impact?

Weak answer: "Our engineers assess whether it's a significant change."

Strong answer: "Every change goes through our CCB where our DQE applies a standard assessment. The DQE asks: 'Does this change affect clinical workflow or patient safety?' Based on the answer and supporting analysis, the change is classified HIGH, MEDIUM, or LOW. Classification determines rigor level. Here's an example CCB record showing the assessment criteria applied."

MOCK INSPECTION Q&A: What happens if an urgent security patch is needed?

Weak answer: "We have an expedited process that bypasses normal review."

Strong answer: "We have an expedited process that accelerates review but doesn't bypass it. Security patches still go through CCB—we can convene CCB within hours for critical issues. The classification still happens, but the timeline is compressed. Here's an example of a critical CVE response: CCB met within 4 hours of identification, classified as MEDIUM because the vulnerability didn't affect clinical functionality, and the fix was deployed within 48 hours with full evidence trail."

G.4 Verification Questions

MOCK INSPECTION Q&A: How do you determine test coverage is adequate?

Weak answer: "We test all the requirements."

Strong answer: "We use multiple coverage criteria: requirements coverage (every requirement has at least one test), risk coverage (every risk control has verification), code coverage (measured by automated tools with thresholds), and boundary coverage (edge cases explicitly tested). Here's our coverage analysis for this release showing each metric. We also do independent verification review where DQE confirms adequacy before approving."

MOCK INSPECTION Q&A: Show me the test protocol for this feature.

Weak answer: "Let me find that document."

Strong answer: "Here's the test protocol for feature X. It includes: test objectives tied to requirements, preconditions and setup, test steps with expected results, actual results documented during execution, and pass/fail determination. The protocol was executed on March 12th—here are the results. DQE approved verification on March 12th at 2:15 PM—here's the approval record."

G.5 Post-Market Questions

MOCK INSPECTION Q&A: How do complaints feed back into design?

Weak answer: "We have a complaint handling process."

Strong answer: "Complaints enter our design control system through Intake, just like any other signal. When PMS Manager identifies a trend or safety concern, they create a signal that goes to CCB. The signal is classified like any other change—if design modification is warranted, it enters a sprint. Here's an example: these three complaints about timeout behavior were identified as a trend, entered Intake, classified as MEDIUM at CCB, and resulted in the fix we discussed earlier. The loop is closed—here's the effectiveness monitoring showing no recurrence."

MOCK INSPECTION Q&A: How do you update your risk analysis based on field experience?

Weak answer: "We review complaints periodically."

Strong answer: "Field experience updates our risk file in two ways. First, complaints provide data for P1 estimates—if a hazardous situation occurred in the field, that's evidence affecting probability estimates. Second, trends may reveal new hazardous situations not previously identified. Here's an example: field complaints revealed this use pattern we hadn't anticipated. We added it as a new hazardous situation, assessed the risk, and implemented a control. The risk file history shows the update dated within two weeks of trend identification."

Appendix H — Complete Walkthrough: A HIGH Risk Feature

This appendix provides a walkthrough of a HIGH risk feature from intake to release, demonstrating the full rigor path.

Figure H-1: FEAT-892 HIGH Risk Feature - Critical Alert Implementation

EPIC OVERVIEW

Epic ID:	FEAT-892
Title:	Critical Alert Feature
User Story:	As a clinician, I want to be alerted when risk scores exceed threshold
Safety_Impact:	HIGH
CCB_Reference:	CCB-2026-032
CCB_Rationale:	New safety-critical alerting system - affects clinical decision timing
RA_Consultation	Required - potential submission impact reviewed

DECOMPOSED STORIES

Story ID	Title	Safety_Impact	Hazard Link
FEAT-892-1	Threshold configuration UI	MEDIUM	HA-0044
FEAT-892-2	Alert trigger logic	HIGH	HA-0044, HA-0045
FEAT-892-3	Visual alert indicator	HIGH	HA-0046
FEAT-892-4	Audio alert tone	HIGH	HA-0046, HA-0047
FEAT-892-5	Alert acknowledgment workflow	HIGH	HA-0048
FEAT-892-6	Alert history logging	MEDIUM	HA-0049

Figure H-1: FEAT-892 HIGH Risk Feature — Critical Alert Implementation

Scenario

Product wants to add a new 'Critical Alert' feature that highlights risk scores above a configurable threshold with visual and audio alerts. This affects clinical decision-making directly by determining which patients receive priority attention.

Week Before Sprint: Intake & CCB

Product Owner Lisa creates Epic FEAT-892: 'Critical Alert Feature'

• User story: 'As a clinician, I want to be alerted when risk scores exceed my configured threshold so I can prioritize critical patients'

• Lisa adds to Intake queue for CCB review

CCB convenes with DQE Maria, PM Alex, Engineering Lead Tom, Product Owner Lisa, RA consultant James (required for potential HIGH items).

DQE Maria applies the universal question: 'Does this affect clinical workflow or patient safety?'

Analysis discussion:

- Alert threshold directly affects which patients get attention
- Missed alert = potential delayed treatment
- False alert = alert fatigue, desensitization, distraction
- This is clinical decision support modification

DQE Maria classifies as HIGH RISK:

- Directly affects clinical prioritization decisions
- New hazardous situations possible (missed alerts, false alerts)
- Full rigor required: DQE at all stages, RA consulted, dual approval at release

⚠ CCB CLASSIFICATION RATIONALE (causal chain form — required for HIGH)

Hazard: Alert triggering logic / threshold configuration UI — device characteristic that determines which patients receive attention and when.

Hazardous Situation A (missed alert): Clinician not alerted when patient risk score exceeds threshold; clinician exposed to unrecognized deterioration.

P1 basis: No existing alert mechanism; every above-threshold event is an exposure.

Hazardous Situation B (false alert): Clinician alerted for below-threshold scores; alert fatigue creates desensitization; future real alerts may be ignored.

P1 basis: Configuration UI permits clinically inappropriate thresholds; no validation.

Harm A (missed alert): Delayed or missed clinical intervention; patient deterioration or adverse outcome.

P2 basis (Situation A): Clinical literature supports high harm probability for missed critical alerts in deteriorating patients. P2 is driven by whether the clinician takes appropriate action after exposure; absence of the alert removes the clinical decision point entirely. RA to confirm against predicate before release.

Harm B (false alert / fatigue): Clinician desensitized by repeated false alerts fails to respond to a subsequent real alert; patient deterioration or adverse outcome from delayed intervention.

P2 basis (Situation B): Alert fatigue and habituation are clinically documented; probability of harm given repeated false-alert exposure is non-trivial. At CCB time (before control selection), no controls yet exist for Situation B — P2 residual risk is high pending control definition and verification. Control selection in Days 3–5 must explicitly address which mechanism reduces P1 (exposure frequency) vs P2 (harm given exposure) for this situation. RA to confirm whether predicate includes fatigue mitigation evidence and whether additional controls beyond threshold validation are required before release.

Safety_Impact = HIGH. Complete ISO 14971 analysis in risk file (HA-0045/46/47).

RA James notes: 'This may affect our 510(k) predicate comparison. I need to review before release.'

Sprint Planning

HIGH risk item requirements for Sprint 48:

- DQE Maria attends all relevant standups
- Risk file must be updated before implementation completes
- RA James consulted on any changes to alert logic
- Verification requires DQE approval
- Release requires DQE + RA dual approval

Sprint commitment includes FEAT-892 plus supporting stories:

- STORY-893: Alert threshold configuration UI
- STORY-894: Alert triggering logic
- STORY-895: Visual alert display
- STORY-896: Audio alert integration

Each story linked to parent Epic with HIGH classification inherited.

Days 1-2: Design & Initial Risk Analysis

ENGINEERING: Architecture design, algorithm definition, edge case identification

PRODUCT: Refines SRS section 5.1 (Alert Requirements) with specific threshold ranges, alert behaviors, and configuration options

DQE: Creates new hazard entries:

- HA-0045: Alert fails to trigger when threshold exceeded (missed alert)
- HA-0046: Alert triggers incorrectly when threshold not exceeded (false positive)
- HA-0047: Alert configuration allows clinically inappropriate thresholds

DQE documents hazardous situations and preliminary severity estimates for each hazard.

Days 3-5: Implementation & Risk Control Definition

ENGINEERING: Core alert logic implemented, configuration UI completed

PRODUCT: Reviews implementation against SRS, adds clarifications for boundary behaviors

DQE identifies and documents risk controls:

• RC-0045: Threshold validation with min/max bounds based on clinical guidelines

• RC-0046: Alert acknowledgment logging — post-market detectability measure. Captures acknowledgment timestamps and override events; enables rapid signal detection if usage patterns suggest elevated residual risk. Not a direct P2 reducer unless escalation on non-acknowledgment is implemented and verified.

• RC-0047: Default threshold from published clinical guidelines with documented rationale

DQE updates P1/P2 estimates based on design decisions and consultation with clinical team. Control-to-probability mapping: RC-0045 and RC-0047 reduce P1 (Situation A: fewer missed alerts; Situation B: fewer clinically inappropriate threshold exposures). No control in this scope directly reduces P2 for either situation — harm probability given exposure is governed by clinical response patterns, not by these controls. RC-0046 provides post-market detectability; it is not a P2 reducer. This residual P2 position is documented in the risk file (HA-0045/46/47) and accepted in the benefit-risk conclusion at release.

Days 6-8: Integration & Test Planning

ENGINEERING: Integration complete, internal testing underway

PRODUCT: Final SRS review, confirms accuracy against implementation

DQE: Defines verification requirements for each risk control, reviews test approach with Engineering

Test cases defined:

• TC-FEAT892-01: Verify alert triggers at exactly threshold (RC-0045)

• TC-FEAT892-02: Verify alert does NOT trigger below threshold

• TC-FEAT892-03: Verify threshold validation rejects out-of-range values (RC-0045)

• TC-FEAT892-04: Verify alert acknowledgment is logged (RC-0046)

• TC-FEAT892-05: Verify default threshold matches clinical guideline value (RC-0047)

Days 9-10: Verification & Approval

DQE Maria witnesses critical tests TC-01 through TC-03 (required for HIGH risk).

Test Results:

• TC-FEAT892-01: PASS - Alert triggers at threshold boundary

• TC-FEAT892-02: PASS - No alert below threshold

• TC-FEAT892-03: PASS - Out-of-range configuration rejected with message

• TC-FEAT892-04: PASS - Acknowledgment logged with timestamp and user ID

• TC-FEAT892-05: PASS - Default matches guideline value

DQE approves verification evidence.

Definition of Done confirmed: Engineering ✓, Product ✓, DQE ✓

Release Gate

Release Gate meeting for Sprint 48 release:

Checklist for FEAT-892 (HIGH RISK):

✓ SRS Status: Approved (Lisa signed)

✓ Risk Status: Approved (Maria signed, includes new hazards and controls)

✓ Test Status: All PASS, witnessed

✓ Evidence Links: Complete

✓ Traceability: FEAT-892 → SRS 5.1 → HA-0045/46/47 → RC-0045/46/47 → TC-*

DQE Maria: 'Evidence is complete. Risk file is current with new hazards and verified controls. Verification demonstrates controls are effective. APPROVED.'

✓ RELEASE DECISION — BENEFIT-RISK CONCLUSION (required for HIGH)

Severity: HIGH. Worst plausible harm: missed critical alert leading to delayed intervention in deteriorating patient. **Most likely harm:** alert delay with clinician-initiated recovery via manual retry.

P1 (exposure): Addressed by RC-0045 (threshold validation) and RC-0047 (default from clinical guidelines). Residual P1 reflects only borderline cases.

P2 (harm given exposure): P2 is driven by whether the clinician takes appropriate action after being exposed to the alert/output. The P2 basis is the clinical rationale that acting on the alert prevents or mitigates the specified harm, supported by objective evidence that the alert/output content is clinically appropriate and presented at the point of decision. RC-0046 (acknowledgment logging) does not, by itself, reduce P2; it provides objective evidence of acknowledgments/overrides and enables post-release monitoring and rapid corrective action if usage patterns suggest elevated residual risk. Clinical guideline-based defaults primarily affect P1 (exposure) by reducing false alerts / unnecessary exposures. RA confirmed the change remains within the scope of current clearance.

P2-reducing control(s) (if any): [Identify the specific control(s) that reduce harm given exposure; do not list logging/auditability unless it directly enables timely clinical intervention.]

Benefit-risk conclusion: Benefit of real-time critical alert for deteriorating patients substantially outweighs residual risk with validated controls in place.

Residual risk accepted. Release approved.

RA James: 'I've reviewed against predicate. The alert feature is within the scope of our current clearance as a visualization enhancement. No new submission required. Labeling update needed for user manual. APPROVED.'

Dual approval achieved. Release proceeds.

HIGH vs MEDIUM Rigor Comparison

The two walkthroughs above demonstrate the same framework operating at different rigor levels. The difference is not in the process—both follow the identical CCB → Sprint → Verification → Release Gate sequence. The difference is in the depth of involvement at each stage. Table H-1 makes this contrast explicit.

Table H-1: HIGH vs MEDIUM Rigor Comparison

Aspect	MEDIUM (Bug example)	HIGH (Feature example)
DQE involvement	Key checkpoints	All stages, daily standup
RA involvement	Informed	Consulted, approval required
New hazard entries	No (existing applied)	Yes (3 new hazards)
Risk control definition	Existing controls applied	New controls defined
Test witnessing	Not required	DQE witnesses critical tests
Release approval	DQE only	DQE + RA dual approval
Submission review	No	Yes (RA confirms no new submission)

The critical insight is that the rigor difference is proportional, not binary. A MEDIUM item is not "uncontrolled"—it still has DQE oversight, verification approval, and traceable evidence. It simply does not require the daily involvement, test witnessing, and dual approval that a HIGH item demands. This proportionality is what makes the framework sustainable: teams can move quickly on LOW and MEDIUM items while applying full rigor exactly where patient safety requires it.

Notice that the escalation points are concrete and auditable. An inspector reviewing a HIGH item can verify that DQE witnessed testing and that RA provided independent approval. An inspector reviewing a MEDIUM item can verify that DQE reviewed and approved evidence. In both cases, the evidence trail demonstrates that rigor matched risk—which is exactly what 21 CFR 820.30 and ISO 14971 require.

Appendix I — EU MDR Compliance Mapping

This framework is primarily designed for FDA compliance but maps directly to EU MDR requirements. For comprehensive EU MDR/IVDR coverage including post-market surveillance obligations (Articles 83-86), IVDR-specific considerations, and Notified Body audit alignment, see Appendix LLL.

I.1 Technical Documentation (Annex II)

EU MDR Annex II defines the structure of technical documentation that manufacturers must compile and maintain. Unlike the FDA's Design History File—which has no prescribed internal structure—Annex II organizes documentation into six numbered sections, each with specific content expectations. Table I-1 maps each Annex II section to the corresponding framework element. The mapping demonstrates that sprint-generated evidence satisfies EU MDR documentation requirements without creating a parallel documentation system. The key architectural point: Sprint DHF packets compile directly into Annex II sections. A team running this framework for FDA compliance is simultaneously generating the evidence EU MDR requires—the difference is packaging, not content.

Table I-1: EU MDR Technical Documentation Mapping

MDR Annex II Section	Framework Element
1. Device description & specification	SRS, intended use documentation
2. Information supplied by manufacturer	Labeling, IFU (RA responsibility)
3. Design and manufacturing info	SDS, Sprint DHF packets, verification evidence
4. General safety & performance	Risk file, usability engineering file
5. Benefit-risk analysis	Overall residual risk evaluation
6. Product verification & validation	Test protocols, execution records, clinical evaluation

Two sections deserve special attention. Section 4 (General Safety and Performance Requirements) requires explicit mapping of each applicable GSPR from Annex I to the evidence that demonstrates compliance. The risk file and usability engineering file generated during sprints provide this evidence, but RA must maintain a GSPR checklist that cross-references specific Hazard_Register entries and verification results to individual GSPRs. Section 6 (Product Verification and Validation) must include clinical evaluation—even for software-only devices. For SaMD, clinical evaluation may reference clinical literature, performance studies, or real-world performance data rather than traditional clinical trials, but the evaluation must exist and must be current. See Appendix LLL for comprehensive EU MDR/IVDR implementation guidance.

I.2 Quality Management System (Article 10)

MDR Article 10 requires a QMS covering design control. The framework addresses these requirements:

Design control procedures: Documented through this framework's SOPs covering CCB, sprint execution, verification, and release.

Risk management: ISO 14971 implementation through the risk file, CCB classification, and risk control verification.

Post-market surveillance: PMS integration through Intake, trend analysis, and closed-loop corrective action.

I.3 Unique Device Identification (Article 27)

The Sprint DHF packet should include:

- UDI assignment for the release (if new UDI required)
- UDI-DI and UDI-PI tracking
- EUDAMED registration status

I.4 Key Differences from FDA

While the framework's sprint mechanics work identically for FDA and EU MDR compliance, the regulatory expectations differ in important ways. Table I-2 highlights the structural differences that affect how teams organize their evidence and documentation.

Table I-2: FDA vs EU MDR Key Differences

Aspect	FDA (21 CFR 820)	EU MDR (2017/745)
Technical documentation	DHF (flexible format)	Annex II structure required
Risk management	Referenced, no specific standard	ISO 14971 explicitly required
Clinical evidence	Varies by device	Clinical evaluation always required
Post-market	MDRs, complaints	PMS system per Article 83-86
Person responsible	Not specified	PRRC required (Article 15)

The practical impact of these differences falls primarily on two roles. RA must manage the structural requirements of Annex II technical documentation, which demands a more prescriptive format than the FDA's flexible DHF approach. The PMS Manager faces significantly expanded obligations under EU MDR Articles 83–86, including mandatory PMS Plans, periodic safety update reports (PSURs), and structured trend reporting that goes beyond FDA's complaint-driven model.

For the sprint team, these differences are largely invisible. The CCB still classifies changes. DQE still owns verification. Evidence is still captured during the sprint. The framework's core mechanics do not change—what changes is how that evidence is packaged for submission and how post-market signals are processed. Organizations operating in both markets can use a single sprint governance process and branch only at the documentation and submission layers. For the full EU MDR/IVDR integration including Notified Body audit alignment, see Appendix LLL.

Appendix J — Complete Walkthrough: Complaint to Resolution

This appendix demonstrates the complete closed-loop process from field complaint through design change and effectiveness verification.

Day 0: Complaint Reception

Customer support receives call from Memorial Hospital: 'Your software froze during a patient assessment. We had to restart and re-enter all the data. This has happened three times this month.'

Customer Support Rep creates complaint record COMP-2024-0156:

• Reported issue: System freeze during patient assessment

• Frequency: 3 occurrences in past month at this site

• Impact: Data loss, workflow disruption, delayed patient care

• Patient harm reported: No (delays but no adverse outcomes)

Day 1: Complaint Triage

Complaints Specialist reviews COMP-2024-0156:

• Is this a complaint per 21 CFR 803.3? YES - malfunction of device

• Is this reportable? Needs PMS Manager assessment

• Complaint logged in complaint database with required fields

PMS Manager Jennifer reviews:

• Similar complaints at other sites? Running query...

• Query returns: 7 similar complaints across 4 sites in past 60 days

• Trend identified: System freeze pattern, appears correlated with large datasets

Day 2: Trend Analysis & Signal Entry

PMS Manager Jennifer documents trend analysis:

• 7 complaints with similar freeze pattern

• All occurred during patient assessment workflow

- All sites had dataset sizes >10,000 records
- No patient harm reported, but workflow impact significant
- Root cause unknown - engineering investigation needed

Jennifer creates Intake signal: 'SIGNAL-2024-0089: System freeze trend during high-volume assessments - investigation required'

Signal enters CCB queue for classification.

Day 3: CCB Classification

CCB convenes. Attendees: DQE Maria, PM Alex, Engineering Lead Tom, PMS Manager Jennifer.

Jennifer presents the trend analysis and complaint pattern.

DQE Maria applies universal question: 'Does this affect clinical workflow or patient safety?'

- System freeze interrupts clinical assessment - YES
- Data loss requires re-entry - workflow impact - YES
- Could delayed assessment affect patient? Potentially - YES

DQE Maria classifies as MEDIUM RISK:

- Not HIGH because no patient harm occurred and freeze is recoverable
- Not LOW because clinical workflow is directly affected
- Investigation sprint authorized

Signal assigned to Sprint 49 for root cause analysis.

▸ CCB CLASSIFICATION RATIONALE — SIGNAL-2024-0089 (causal chain form — required for MEDIUM)

Hazard: System memory management under high-volume data load — device characteristic: software loads entire dataset into memory without pagination or load threshold, creating potential for unresponsive state during clinical assessment.

Hazardous Situation: Clinician initiates patient assessment on a site with dataset >10,000 records; system freezes mid-workflow; clinician is exposed to data loss and workflow interruption at a clinical decision point.

P1 basis: 7 confirmed complaints across 4 sites in 60 days; all correlated with dataset size >10,000 records. P1 updated upward from baseline — this is new objective field evidence (complaint data) superseding any prior pre-release estimate. Signal updates HA-0048 P1 entry. (Note: prior P1 was lab-based; field rate may be higher.)

Harm: Delayed or incomplete patient assessment due to system freeze and data re-entry burden (i.e., delayed clinical intervention in time-critical cases).

Most likely harm: workflow disruption and data re-entry; assessment restarted with no patient injury.

Worst plausible harm: freeze during time-critical assessment leads to delayed clinical intervention for deteriorating patient.

P2 basis: System freeze is visible and recoverable — clinician can restart and re-enter data. No harm occurred in any of the 7 reported events. P2 is low to moderate: harm requires both an unrecovered freeze AND time-critical clinical context with no alternative access pathway. Severity does not change from prior estimate; P1 (exposure frequency) is what the signal updates.

Safety_Impact = MEDIUM. Risk file updated: HA-0048 (System Unresponsive During Data Loading) — P1 revised upward based on complaint data. P2 and severity unchanged. RC-0048 (paginated loading) to be defined and verified in Sprint 50.

Causal discipline note: *High patient volume, large dataset size, and operational load conditions are environmental conditions that surface this hazardous situation — they are not the hazard. The hazard remains the device's architectural characteristic: software that loads entire datasets without pagination or memory threshold. Environmental conditions go in the Hazardous Situation; device characteristics go in the Hazard. The same discipline applies when the contributing condition is a network failure, an AWS outage, or any other operational context.*

Sprint 49: Investigation

Engineering investigates root cause:

- Day 1-2: Reproduce issue with large datasets
- Day 3: Root cause identified - memory management issue when loading >10,000 records
- Day 4: Design fix approach: Implement pagination, add memory monitoring
- Day 5: Fix designed, ready for implementation sprint

Investigation findings documented:

- Root cause: Single query loads entire dataset into memory
- Trigger condition: Dataset >10,000 records
- Fix approach: Paginated loading with progress indicator
- Estimated fix: 5 story points in Sprint 50

DQE updates risk file:

- HA-0048: System unresponsive during data loading (existing hazard, new root cause identified)
- P1 updated based on complaint frequency data
- RC-0048: Paginated loading with memory threshold (proposed)

Sprint 50: Implementation

Fix implementation follows MEDIUM rigor path:

- Day 1-3: Implementation of paginated loading
- Day 4-5: Memory monitoring and graceful degradation
- Day 6-8: Integration testing with large datasets
- Day 9-10: Verification with DQE approval

Test cases verify fix:

- TC-FIX0048-01: Load 5,000 records - no freeze - PASS
- TC-FIX0048-02: Load 15,000 records - paginated, no freeze - PASS
- TC-FIX0048-03: Load 50,000 records - paginated with progress, no freeze - PASS

• TC-FIX0048-04: Memory stays below threshold during loading - PASS

DQE approves verification. Definition of Done confirmed.

Release & Deployment

Release Gate confirms all evidence complete. Fix deployed in Release 4.2.1.

Affected sites notified of update. Deployment tracking:

• Memorial Hospital: Updated Day 1 post-release

• Other 3 affected sites: Updated within 2 weeks

• Remaining sites: Updated per normal rollout schedule

Effectiveness Monitoring

PMS Manager Jennifer monitors for recurrence:

• 30-day post-deployment: 0 freeze complaints from updated sites

• 60-day post-deployment: 0 freeze complaints from updated sites

• 90-day post-deployment: 0 freeze complaints, 2 complaints from non-updated sites (expected)

Effectiveness criteria met: Zero recurrence at updated sites for 90 days.

CAPA Closure

CAPA record CAPA-2024-0023 closed:

• Root cause identified: Memory management issue ✓

• Corrective action implemented: Paginated loading ✓

• Verification evidence: Test results attached ✓

• Effectiveness demonstrated: 90-day monitoring complete ✓

• Risk file updated: HA-0048 and RC-0048 documented ✓

Complete traceability chain:

COMP-2024-0156 → SIGNAL-2024-0089 → CAPA-2024-0023 → FIX-0048 → Release 4.2.1 → Effectiveness Verified

> **KEY INSIGHT:** This is what a closed loop looks like: complaint → trend → investigation → fix → verification → deployment → effectiveness. Every step traceable, every transition documented.

Appendix K — Engineering Leader Quick Reference

This appendix collects key insights for engineering leaders who need to understand the framework quickly without reading every chapter.

K.1 The Sprint as Compliance Unit

FOR ENGINEERING LEADERS: What This Means for Your Team

Every sprint produces a compliance-ready packet. You're not doing extra work—you're organizing existing work to be audit-ready. Think of it this way: the Jira ticket IS the design history record. The code review IS part of design review. The test results ARE verification evidence. You already do this work; now you're capturing it in a form regulators recognize.

Table K-2: Sprint Activity to Design Control Mapping

Sprint Activity	Design Control Term	What Changes
Story refinement	Design input	Link to SRS section
Technical design discussion	Design review	Document decision rationale
Code review	Design output review	No change (already documented)
Unit tests	Unit verification	Track coverage metrics
Integration tests	Integration verification	Link to requirements
QA testing	System verification	DQE approval for HIGH/MEDIUM

K.2 The Tracker is Your Friend

FOR ENGINEERING LEADERS: Visibility Without Surveillance

The Design Control Tracker isn't surveillance—it's visibility. When leadership can see real status in real time, they stop asking for status reports. When blockers are visible, they get resolved faster.

When evidence is tracked as it's created, there's no scramble at release time. Your team benefits from transparency: no surprises, no blame games, no last-minute documentation marathons.

K.3 Jira is Your Design History File

FOR ENGINEERING LEADERS: Five Fields That Change Everything

Add these five custom fields to every ticket: Safety_Impact (CCB classification), CCB_Reference (link to CCB record), SRS_Link (requirement traceability), Risk_Link (hazard traceability), and Verification_Status (DQE approval). With these fields populated, your Jira ticket IS your design history record. The DHF isn't a separate document—it's a view of your Jira data.

K.4 Who to Call for What

Table K-1: Quick Reference — Who to Contact

Question	Go To	Not To
Is this HIGH or MEDIUM risk?	DQE	Engineering Lead
Can we skip CCB for this?	PM (answer: no)	Anyone else
Does this affect our 510(k)?	RA	DQE or PM
Is verification adequate?	DQE	Engineering Lead
Can we release Friday?	Tracker (then PM)	Engineering Lead
Who owns this blocker?	Tracker	Meeting discussion

K.5 CCB is Not Your Enemy

FOR ENGINEERING LEADERS: Classification, Not Approval

CCB doesn't slow you down if you understand its purpose. CCB classifies risk—it doesn't approve work. Work proceeds regardless of classification. A LOW item proceeds with minimal overhead. A HIGH item proceeds with more checkpoints. You're not asking permission; you're getting routing. The 2-minute classification at CCB saves the 2-week retrofit scramble at release.

K.6 DQE is Your Quality Partner

FOR ENGINEERING LEADERS: Embedded, Not External

The DQE role exists to operate at your speed. Traditional QA operates at audit cadence—they check your work after you're done. DQE operates at sprint cadence—they participate in your work as it happens. A good DQE makes your life easier: they catch risk issues before they become blockers, they approve verification while context is fresh, and they ensure releases don't stall on documentation.

K.7 What 'Done' Really Means

Your team's Definition of Done now requires three sign-offs:

- Engineering: Code complete, reviewed, tests passing
- Product: Requirements linked, SRS current
- DQE: Verification approved, traceability complete (for MEDIUM/HIGH)

Items that skip any sign-off are NOT done. The Jira workflow enforces this. This isn't bureaucracy—it's prevention. Every item that ships with incomplete evidence creates audit risk that compounds. Complete evidence now is always cheaper than retrofit later.

Appendix L — Chapter Summaries for Quick Reference

This appendix provides condensed summaries of each chapter for quick review.

Part I: The Problem

Chapter 1: The Pain Every SaMD Team Knows

The retrofit tax costs 20-40% of release effort. Symptoms include: documentation lag, release delays, QA-engineering friction, and 'engineering-only' exceptions. The pain is real and measurable.

Chapter 2: Why Agile and Compliance Feel Like Enemies

The conflict is organizational, not inherent. Regulators don't require sequential phases or locked requirements. Organizations created constraints that regulations don't mandate.

Chapter 3: The Inspection That Reveals Everything

Timestamps tell the truth. Evidence created during development demonstrates understanding; evidence created afterward demonstrates compliance effort. Inspectors can tell the difference.

Part II: The Insight

Chapter 4: The Temporal Insight

The conflict is temporal, not philosophical. Before = waterfall dysfunction. After = retrofit tax. During = native design control. Same work, different timing, different outcomes.

Chapter 5: Design Control Was Never the Enemy

Regulations describe what must be done, not when. 21 CFR 820.30, ISO 14971, and IEC 62304 all allow iterative development. AAMI TIR45 makes this explicit.

Part III: The Foundation

Chapter 6: The Language That Changes Everything

Hazard → Hazardous Situation → Harm. Hazards are device characteristics, never user actions. Terminology locks prevent drift. P1 is exposure probability; P2 is harm probability given exposure.

Chapter 7: Who Does What

DQE operates at sprint cadence; RA at milestone cadence. Clear role separation prevents bottlenecks. DQE owns classification; they don't take a vote. RACI clarity enables speed.

Part IV: The Framework

Chapter 8: The Regulated Sprint

Intake → CCB → Sprint → Release → Field → PMS → Intake. The loop never ends. The sprint is the smallest auditable unit. Evidence is captured during, not after.

Chapter 9: The Risk Gate in Depth

CCB is classification, not approval. The universal question: 'Does this affect clinical workflow or patient safety?' Classification determines rigor, not permission.

Chapter 10: Parallel Execution

Requirements, risk, design, and testing evolve together. Timestamps should distribute across the sprint. Clustering reveals reconstruction.

Chapter 11: The Design Control Tracker

Real-time visibility into compliance status. Release readiness is calculated, not assessed. The tracker must be the single source of truth.

Chapter 12: Integrating Your Tools

Five custom fields transform Jira into a design history system. Traceability is achieved through links, not matrices. Sprint DHF packets generate from Jira data.

Part V: Making It Work

Chapter 13: Usability and Use-Related Risk

Usability feeds into risk management. User behavior appears in hazardous situations. Formative findings flow into sprints; summative findings verify control effectiveness.

Chapter 14: Post-Market Surveillance

Complaints are design input. PMS signals enter Intake like any other change. Closing the loop means tracing from complaint to CAPA to fix to effectiveness. For EU-marketed devices, PMS Manager owns PMS Plan (Article 84) and PSUR preparation (Article 86).

Chapter 15: Release Gates and Definition of Done

Three sign-offs required: Engineering, Product, DQE. Release Gate confirms evidence; it doesn't evaluate work. Validation sprints freeze code while validation completes.

Chapter 16: AI, Cybersecurity, and Emerging Signals

All signals flow through the same process. PCCP boundaries define pre-approved AI changes. CVE severity ≠ safety impact.

Part VI: Inspection Defense

Chapter 17: Facing the Inspector

Lead with narrative, not documents. Let timestamps tell the story. Answer questions directly with specific evidence. For Notified Body audits, emphasize Technical Documentation structure and GSPR traceability. When you don't know, say so.

Chapter 18: Common Failure Modes

Watch for: CCB rubber-stamping, engineering-only exceptions, evidence reconstruction, tracker gaming, CCB bottleneck, role confusion. Each has detection methods and fixes.

Part VII: The Future

Chapter 19: Where This Goes From Here

The framework evolves with the industry. Continuous learning systems will accelerate the loop. The goal is invisible compliance—embedded so deeply that no one thinks about it.

Appendix M — Sprint DHF Packet Template and Contents

This appendix provides detailed guidance on Sprint DHF Packet structure and contents.

M.1 Folder Structure

Sprint_XX_DHF/

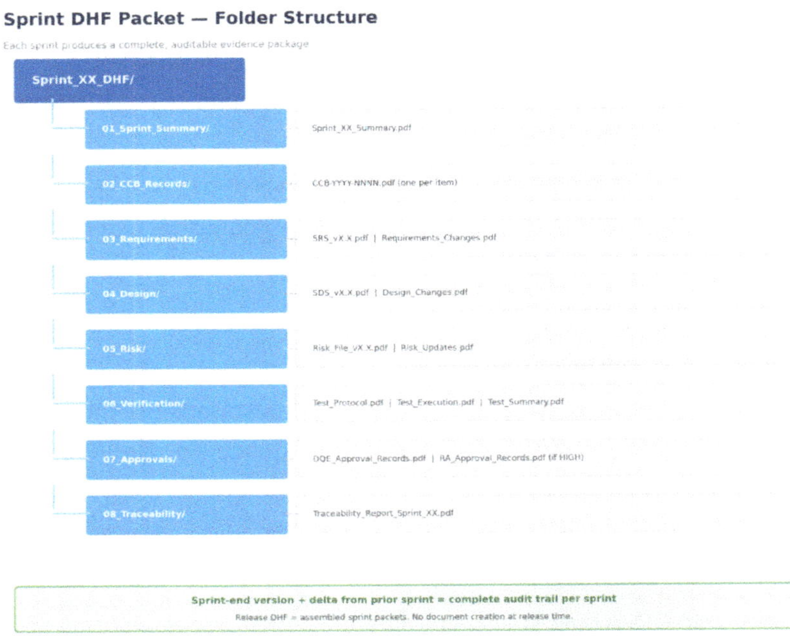

Figure M-1: Sprint DHF Packet Folder Structure — Self-Contained Evidence Package per Sprint

M.2 Sprint Summary Contents

The Sprint Summary document includes:

- Sprint number and dates
- List of items completed with classifications
- Risk distribution (HIGH/MEDIUM/LOW counts)

- Key decisions and rationale
- Carryover items (if any)
- Sprint metrics (velocity, evidence completion rate)

M.3 Minimum Contents by Risk Level

HIGH Risk Items

□ CCB classification record with rationale

□ SRS section (approved, signed)

□ Risk file update (approved, includes new hazards/controls)

□ SDS section if architecture affected (approved)

□ Test execution evidence with DQE witnessing record

□ DQE approval record

□ RA review record

□ Full traceability: Item → SRS → Hazards → Controls → Tests

MEDIUM Risk Items

□ CCB classification record

□ SRS linkage

□ Risk file update

□ Test execution evidence

□ DQE approval record

□ Traceability: Item → SRS → Hazards → Tests

LOW Risk Items

□ CCB classification record

□ Test execution evidence

□ Basic traceability

M.4 Traceability Report

The Traceability Report shows relationships between:

- Sprint items → SRS requirements
- SRS requirements → Hazards
- Hazards → Risk controls
- Risk controls → Test cases
- Test cases → Execution results

This report generates from Jira link data. It demonstrates that every sprint item connects to requirements, requirements connect to risk, and risk connects to verification.

Appendix N — Tool Validation Guidance

This appendix addresses validation requirements for tools used in the framework.

N.1 Jira Validation

Jira, when configured as described in this framework, requires validation as a quality system tool. The validation should demonstrate:

Intended use: Work item tracking, status tracking, traceability, evidence storage

Risk assessment: What happens if Jira fails? Mitigation: backups, export capability, alternative access

Functional requirements for custom fields:

- Safety_Impact field accepts only HIGH/MEDIUM/LOW
- Workflow prevents transitions without required fields
- Links maintain referential integrity
- Timestamps are accurate and immutable
- Audit trail captures all changes

Validation approach: IQ (installation qualification), OQ (operational qualification), PQ (performance qualification). For cloud-hosted Jira, vendor qualification may substitute for IQ.

N.2 Tracker Validation

If the Design Control Tracker is a separate tool (Excel, database, etc.), it requires:

- Accuracy verification: Status matches source data
- Calculation verification: Release readiness logic correct
- Data integrity: Cannot be modified without audit trail
- Backup/recovery: Data can be restored if lost

N.3 Document Control

Document control system (for SRS, Risk File, etc.) requires:

- Version control: Every change creates new version
- Access control: Appropriate edit permissions
- Electronic signatures: 21 CFR Part 11 compliance if electronic
- Retention: Documents retained per retention schedule

REGULATORY ANCHOR: 21 CFR Part 11 applies to electronic records and signatures. If SRS approvals, risk file approvals, and verification approvals are electronic, the system must meet Part 11 requirements for authenticity, integrity, and non-repudiation.

N.4 Test Management

Test management tools require:

- Test case version control
- Execution evidence capture (cannot be modified after execution)
- Traceability to requirements
- Status tracking accuracy

Appendix O — SRS Authoring for Regulated Software

This appendix provides detailed guidance on writing software requirements specifications that satisfy both regulatory requirements and practical engineering needs.

O.1 Requirements Structure

Each SRS requirement should include:

Table O-1: SRS Requirement Elements

Element	Purpose	Example
Requirement ID	Unique identifier for traceability	SRS-4.2.1-001
Requirement text	What the software shall do	The system shall display the risk score with two decimal places
Rationale	Why this requirement exists	Precision required for clinical differentiation between adjacent risk levels
Acceptance criteria	How to verify this is met	Display shows '0.00' to '100.00' with no rounding errors at boundaries
Parent ID	Link to user need or system requirement	UN-015, SYS-042
Risk link	Associated hazards	HA-0023

O.2 Common SRS Problems

483 PATTERN: Requirements that don't match implemented behavior. Inspector compares SRS to software and finds discrepancies. The SRS says 'shall display in green' but software displays in blue. Root cause: Requirements not updated when design changed.

483 PATTERN: Requirements that are untestable. Inspector asks 'how do you verify this?' and there's no clear answer. Example: 'The system shall be easy to use.' This cannot be objectively verified. Root cause: Requirements written without verification in mind.

483 PATTERN: Missing requirements for error handling. Software handles errors but SRS doesn't specify error handling behavior. Inspector asks 'where is the requirement for what happens when the sensor disconnects?' Root cause: Happy path focus during requirements.

O.3 Iterative Requirements in Practice

In this framework, requirements evolve during sprints. This is compliant with AAMI TIR45 guidance, but requires discipline:

At sprint start: Initial requirements documented in SRS, marked as 'DRAFT' or 'IN DEVELOPMENT'

During sprint: Product Owner updates requirements as engineering reveals implications. Updates happen in real-time, not batched at sprint end.

At sprint end: Requirements for completed items marked 'APPROVED'. Incomplete items remain 'DRAFT' and carry to next sprint.

Version control: Each sprint produces a new SRS version. Sprint DHF packet references the sprint-end version.

REGULATORY ANCHOR: AAMI TIR45:2023 Section 6.2 states that design inputs may be 'refined iteratively based on evaluation of design outputs.' The key is that requirements are documented before verification completes—not before implementation begins.

O.4 Requirements Traceability

Every requirement should trace in two directions:

Upstream traceability: Requirement traces to user need, market input, or regulatory requirement that justifies its existence. If a requirement cannot justify why it exists, question whether it's needed.

Downstream traceability: Requirement traces to test cases that verify it, design elements that implement it, and risk controls it relates to.

Traceability is maintained through Jira links. The traceability matrix is a query result, not a separate document.

Appendix P — Risk File Structure and Maintenance

This appendix provides detailed guidance on risk file structure and maintenance throughout the product lifecycle.

P.1 Risk File Components

The risk file is not a single document—it's a compilation of risk management records. Components include:

Risk Management Plan: Defines scope, risk acceptability criteria, and risk management process for this product

Hazard Analysis: Systematic identification of hazards and hazardous situations

Risk Estimation: P1, P2, and severity estimates for each hazardous situation

Risk Evaluation: Comparison against acceptance criteria, determination of acceptability

Risk Control: Documentation of controls selected, implemented, and verified

Residual Risk Evaluation: Overall residual risk assessment after all controls implemented

Risk Management Report: Summary of risk management activities and conclusions

P.2 Maintaining Risk During Sprints

The risk file is a living document during development. Each sprint may add, modify, or verify risk content:

New features: May introduce new hazards, new hazardous situations, or new controls

Bug fixes: May verify existing controls, refine P1 estimates, or close risk gaps

Field signals: May add new hazardous situations based on field experience, refine probability estimates

DQE owns risk file updates during sprints. For HIGH risk items, updates should be reviewed by a second qualified person.

P.3 Hazard Identification Checklist

Use this checklist when identifying hazards for new features:

□ Could the software display wrong information? (wrong patient, wrong value, wrong unit)

□ Could the software fail to display critical information? (missing alerts, hidden warnings)

□ Could the software delay critical information? (slow response, queued alerts)

□ Could the software allow unsafe configuration? (dangerous thresholds, conflicting settings)

□ Could the software fail silently? (error without indication)

□ Could the software continue operating with stale data? (sensor disconnection)

□ Could the software allow unintended actions? (accidental confirmation, wrong patient selection)

□ Could the software output be misinterpreted? (ambiguous display, unclear units)

P.4 Risk Control Verification

Every risk control must be verified. Verification approaches by control type:

Table P-1: Risk Control Verification Approaches

Control Type	Verification Approach	Evidence
Input validation	Test with boundary and invalid values	Test execution results
Alerting	Test alert triggers and non-triggers	Test execution with timestamps
Display format	Inspection and usability testing	Screenshots, usability reports
Calculation	Test with known values	Comparison to expected results
Timeout/watchdog	Test timeout behavior	Log files showing timeout handling

Each verification approach in Table P-1 should produce evidence that demonstrates reduction in a specific probability component. Input validation verification (testing boundary and invalid values) produces evidence of P0 reduction — the test results show that erroneous inputs are caught before causing hazardous situations. Alerting verification (testing triggers and non-triggers) produces evidence of P2 reduction — the test results show that hazardous situations are detected before causing harm. Display format verification (inspection and usability testing) produces evidence of Pe reduction — the usability test results show that the redesigned display reduces user interpretation errors.

When documenting risk control verification results in the risk file, explicitly state which probability component the control targets and how the verification evidence supports the post-control probability estimate. A test result showing that input validation rejects 100% of out-of-range values supports a post-control P0 = 0.0 for that specific error mode. A usability test showing that the redesigned alert is noticed by 14 of 15 participants (93%) supports a post-control Pe reduction from the pre-control estimate — but not to zero, because one participant still missed the alert. This level of precision in connecting verification evidence to probability estimates is what distinguishes a defensible risk file from one that simply lists "risk control verified" without analytical substance.

Appendix Q — Verification Testing Best Practices

This appendix provides detailed guidance on verification testing that satisfies regulatory requirements.

Q.1 Test Case Structure

Each test case should include:

Table Q-1: Test Case Elements

Element	Purpose	Example
Test ID	Unique identifier	TC-SRS-4.2.1-001-01
Requirement reference	What this tests	SRS-4.2.1-001
Risk control reference	Risk control verified (if applicable)	RC-0023
Preconditions	Required setup state	System initialized, test user logged in, test patient loaded
Test steps	Actions to perform	1. Navigate to risk display 2. Observe score format
Expected results	What should happen	Score displays as '45.67' with two decimal places
Actual results	What actually happened	(Recorded during execution)
Pass/Fail	Determination	(Recorded during execution)

Q.2 Test Coverage Requirements

Verification should demonstrate coverage across multiple dimensions:

Requirements coverage: Every SRS requirement has at least one test case. Coverage report shows 100% requirements coverage.

Risk control coverage: Every risk control has verification. Coverage report shows 100% control coverage.

Boundary coverage: Boundary values explicitly tested. For numeric inputs, test at minimum, maximum, and mid-range.

Error handling coverage: Error conditions explicitly tested. What happens when inputs are invalid, connections fail, or resources are unavailable?

Regression coverage: Unchanged functionality re-verified. Regression test suite executed for each release.

Q.3 Test Execution Evidence

Test execution must produce evidence that demonstrates:

What was tested: Test case ID and version

When it was tested: Execution timestamp

What software was tested: Software version under test

Who executed: Tester identity

What happened: Actual results observed

What was determined: Pass/fail with justification if fail

For MEDIUM and HIGH risk items, DQE reviews evidence and approves verification. The approval timestamp becomes part of the evidence.

Q.4 Regression Testing Strategy

Regression testing ensures unchanged functionality still works. Strategy should define:

Scope: What's included in regression (typically: all previously verified functionality)

Triggers: When regression runs (typically: every release, after integration)

Automation: Which regression tests are automated vs. manual

Prioritization: If full regression isn't feasible, how to select tests (risk-based)

> **483 PATTERN:** No evidence of regression testing. New features were tested, but unchanged functionality was not re-verified. Inspector asks: 'How do you know the changes didn't break existing functionality?' Organization cannot demonstrate regression coverage.

Appendix R — Validation vs. Verification: When and How

Confusion between validation and verification causes significant compliance problems. This appendix clarifies the distinction and provides practical guidance.

R.1 The Fundamental Difference

VERIFICATION asks: Did we build it right?

- Tests against specifications
- Demonstrates design outputs meet design inputs
- Can happen throughout development
- Typically performed by development/test team

VALIDATION asks: Did we build the right thing?

- Tests against user needs and intended use
- Demonstrates device meets user needs under actual or simulated use conditions
- Typically requires stable release candidate
- Often involves actual users or simulated clinical scenarios

R.2 Verification Activities

Verification happens throughout sprints:

- Unit testing: Code units meet module specifications
- Integration testing: Modules work together as specified
- System testing: Complete system meets SRS requirements
- Risk control verification: Controls function as designed

All verification can complete within the sprint. DQE approval at sprint end confirms verification adequacy.

R.3 Validation Activities

Validation typically requires activities that span beyond a single sprint:

Usability validation (summative): Does the device allow users to accomplish intended tasks safely? Requires stable UI, representative users, clinical scenarios.

Clinical validation: Does the device perform as intended in clinical use? May require clinical studies, field testing, or acceptance testing at customer sites.

Performance validation: Does the device meet performance claims under expected conditions? May require extended operation, load testing, stress testing.

R.4 The Validation Sprint

When validation activities are required, schedule a dedicated Validation Sprint:

Timing: After feature development is complete, before release

Code freeze: No new development on the release candidate during validation

Activities: Summative usability testing, clinical validation, performance validation

Exit criteria: Validation demonstrates device meets user needs and intended use

If validation finds issues: Issues enter Intake, get classified, and if fixes are needed, the cycle restarts. The fixed version requires re-validation.

REGULATORY ANCHOR: 21 CFR 820.30(g) requires design validation 'under actual or simulated use conditions.' Verification testing against SRS is necessary but not sufficient. Validation demonstrates the complete device meets user needs.

Appendix S — Cybersecurity Risk Integration

FDA's cybersecurity guidance requires integration of cybersecurity risks with safety risk management. This appendix provides detailed guidance.

S.1 Cybersecurity as Design Input

Cybersecurity requirements are design inputs just like functional requirements. They should appear in the SRS with the same structure:

Requirement: The system shall require authentication before displaying patient data

Rationale: Prevent unauthorized access to PHI, comply with HIPAA

Acceptance criteria: Login screen requires valid credentials, session timeout after inactivity

Risk link: HA-0055 (unauthorized data access)

S.2 Cybersecurity Hazards

Cybersecurity hazards follow the same structure as safety hazards:

Hazard (device characteristic): System stores patient data without encryption

Hazardous situation: Attacker accesses database, extracts patient records

Harm: Patient privacy violation, potential for discrimination based on health data

Note that cybersecurity hazards can also lead to safety harms:

Hazard: System accepts unvalidated input

Hazardous situation: Attacker injects malicious data that corrupts clinical calculations

Harm: Clinician makes treatment decision based on corrupted data, patient receives wrong treatment

Causal discipline reminder: Environment contributes to hazardous situation; hazard remains the device characteristic (including data handling, logic, and availability design choices). A network outage or attacker action is the environmental condition that creates the hazardous situation — the hazard is the device's failure to protect data or validate input.

S.3 CVE Response Process

When a CVE affects your system:

1. Security team identifies affected component

2. Signal enters Intake: CVE-XXXX-XXXXX with severity and affected versions

3. CCB classifies based on PATIENT SAFETY impact, not just CVE severity

4. Classification guide:

Table S-1: CVE Severity Classification Guide

CVE Severity	Patient Safety Impact	Classification
Critical	Could affect clinical data/functionality	HIGH
Critical	Infrastructure only, no clinical impact	MEDIUM
High	Could affect clinical data/functionality	HIGH
High	Infrastructure only	MEDIUM
Medium/Low	Affects user workflow or data presentation	MEDIUM
Medium/Low	Infrastructure only, no user-facing impact	LOW

HIGH row coloring indicates classification level
MEDIUM row coloring indicates classification level
LOW row coloring indicates classification level

5. Fix proceeds with appropriate rigor based on classification

6. Evidence includes: CVE analysis, classification rationale, fix verification, deployment timeline

> **REGULATORY ANCHOR:** FDA's Cybersecurity Guidance (2023) states: 'Manufacturers should ensure that cybersecurity risks are appropriately addressed as part of the QS regulation's design controls.' The framework

accomplishes this by running cybersecurity signals through the same CCB and risk management process as all other changes.

Appendix T — Detailed CCB Classification Scenarios

This appendix provides extended scenarios demonstrating CCB classification reasoning for various change types.

T.1 Algorithm Changes

Scenario: New Risk Calculation Algorithm

Change: Replace current risk scoring algorithm with machine learning model trained on validated dataset.

CCB Analysis:

• Does this affect clinical workflow? YES - Risk scores guide clinical decisions

• Does this affect patient safety? YES - Different algorithm = different scores = different patient prioritization

• New hazards introduced? YES - Model drift, training bias, edge cases

Classification: HIGH

Rationale: Any change to clinical calculation logic is HIGH by default. Algorithm changes can have non-obvious effects across the score distribution. Full rigor required.

RA Impact: Likely requires regulatory assessment. May need new clinical evidence.

Scenario: Algorithm Performance Optimization

Change: Refactor algorithm implementation for 40% faster execution without changing calculation logic.

CCB Analysis:

• Does this affect clinical workflow? Partially - Faster response improves workflow

• Does this affect patient safety? Analysis needed - Does faster execution change anything?

• Same inputs = same outputs? Verification needed

Classification: MEDIUM

Rationale: If verification confirms identical outputs, MEDIUM is appropriate. The 'same logic, different implementation' pattern requires verification that outputs truly are identical. DQE approval required.

T.2 UI Changes

Scenario: Redesigned Dashboard

Change: Complete visual redesign of clinician dashboard with new layout, colors, and navigation.

CCB Analysis:

• Does this affect clinical workflow? YES - Clinicians must relearn navigation

• Does this affect patient safety? Potentially - Could affect ability to find critical information

• Usability risk? YES - Major UI changes introduce use error potential

Classification: HIGH

Rationale: Major UI redesigns affect how clinicians find and interpret information. Requires formative usability testing during development and may require summative usability validation.

Scenario: Color Scheme Change

Change: Update color scheme from blue/gray to company brand colors (green/white).

CCB Analysis:

• Does this affect clinical workflow? Minimal

• Does this affect patient safety? Need to analyze - Are any colors clinically meaningful?

• Accessibility impact? Need to verify contrast ratios

Classification: MEDIUM (or HIGH if colors are clinically meaningful)

Rationale: Pure cosmetic color changes are MEDIUM. However, if current colors convey clinical meaning (red = critical, green = normal), changing them could cause misinterpretation.

T.3 Infrastructure Changes

Scenario: Database Migration

Change: Migrate from MySQL to PostgreSQL for better scalability.

CCB Analysis:

• Does this affect clinical workflow? No visible change to users

• Does this affect patient safety? Potentially - Data migration could corrupt data

• Migration risk? YES - Data migration is high-risk technical activity

Classification: MEDIUM

Rationale: Infrastructure changes that don't change clinical functionality are MEDIUM. The technical risk is significant but the clinical risk is mitigated by verification of data integrity.

Scenario: Cloud Provider Migration

Change: Move from AWS to Azure hosting.

CCB Analysis:

• Does this affect clinical workflow? No user-visible changes

• Does this affect patient safety? Potentially - Performance, availability, security

• Regulatory considerations? Data residency, security certifications

Classification: MEDIUM (with RA consultation)

Rationale: Cloud migration itself is MEDIUM infrastructure. However, RA should review for data residency and security compliance implications.

T.4 Third-Party Component Updates

Scenario: Security Patch to OS Library

Change: Update OpenSSL from 3.0.1 to 3.0.2 (security patch only).

CCB Analysis:

- Does this affect clinical workflow? No
- Does this affect patient safety? No (unless CVE affects clinical data)
- CVE severity? Check specific CVE for patient safety implications

Classification: LOW (unless CVE has patient safety implications)

Rationale: Security patches to system libraries typically don't affect clinical functionality. Verify through regression testing.

Scenario: Major Version Upgrade of Clinical Library

Change: Upgrade clinical calculations library from v2.x to v3.x (major version with breaking changes).

CCB Analysis:

- Does this affect clinical workflow? Need to analyze
- Does this affect patient safety? HIGH RISK - Clinical calculation library
- Breaking changes? YES - Major version

Classification: HIGH

Rationale: Any change to clinical calculation dependencies requires full verification. Major versions with breaking changes require careful analysis of all calculation impacts.

Appendix U — Release Management Scenarios

This appendix addresses complex release scenarios and their resolution.

U.1 Mixed-Risk Releases

Scenario: Release contains 1 HIGH, 3 MEDIUM, and 15 LOW items.
Release Gate Requirements:
• All 19 items must have complete evidence per their classification
• HIGH item requires DQE + RA dual approval
• MEDIUM items require DQE approval
• LOW items require verification evidence
Practical approach:
1. Prioritize HIGH item completion - it's the blocker for release
2. If HIGH item isn't ready, consider splitting release
3. RA involvement needed for HIGH approval - schedule early
4. Release proceeds only when highest-risk item is complete

U.2 Emergency Hotfix Release

Scenario: Critical bug discovered in production requiring immediate fix.
Process acceleration:
1. Signal enters Intake immediately (don't wait for CCB meeting)
2. Emergency CCB convenes within hours (DQE + PM minimum)
3. Classification determines rigor - even emergencies get classified
4. Compressed sprint (days instead of weeks)
5. Same evidence requirements - compressed timeline
6. Verification must still happen - accelerated, not skipped
7. Release Gate confirms evidence exists
What changes in emergency:
• Meeting timing (hours instead of scheduled)
• Sprint duration (days instead of weeks)
• Parallel activity (more concurrent work)
What does NOT change:
• Classification requirement
• Evidence requirements per classification
• Approval requirements
• Traceability requirements

U.3 Blocked Item Scenarios

Blocked by External Dependency

Scenario: MEDIUM item blocked waiting for third-party API specification.
Options:
1. Remove from release (if not critical)
2. Implement with best-guess specification, plan update when final spec available
3. Hold release (if item is critical)

Decision factors: Clinical importance, release timing, dependency timeline

Blocked by Failing Tests

Scenario: HIGH item implemented but verification fails.
Resolution required before release:
1. Investigate failure - is it implementation bug or test issue?
2. If implementation bug: fix, re-verify (time added)
3. If test issue: fix test, re-execute
4. DQE must approve passing verification before release

Releases do not proceed with failed verification on HIGH/MEDIUM items.

U.4 Release Rollback Scenarios

Scenario: Post-release discovery of significant issue requiring rollback.

Rollback as design change:

1. Signal enters Intake (defect report)
2. CCB classifies severity
3. Rollback decision made (Program Manager)
4. Rollback executed with evidence
5. Customer communication (RA/PMS)
6. Root cause analysis initiates CAPA

Documentation required:

- Reason for rollback
- Approval of rollback decision
- Verification that rollback was successful
- Customer notification record
- CAPA reference for root cause investigation

Appendix V — Design Review Implementation

This appendix clarifies design review requirements and how they integrate with the framework.

V.1 Design Review vs. Code Review

Table V-1: Code Review vs Design Review Comparison

Aspect	Code Review	Design Review
Focus	Implementation quality	Design adequacy
Participants	Engineering peers	Cross-functional (including DQE)
Questions	Is this code correct?	Is this design appropriate?
Timing	Before code merge	At design milestones
Regulatory	Part of good practice	Required by 820.30(e)

Code review is necessary but not sufficient. Design review requires cross-functional perspective that evaluates whether the design meets requirements, addresses risks, and is appropriate for intended use.

V.2 When Design Reviews Are Required

Design reviews are required at 'appropriate stages' per 820.30(e). In this framework:

Before CCB (for HIGH items): Design approach reviewed before classification

During sprint (for HIGH items): Design review at design solidification

At release (for all items): Design review confirms adequacy before release

Design review is implicit in the framework's governance. The three-role Definition of Done (Engineering, Product, DQE) constitutes a design review. However, organizations may want additional formal design reviews for:

- Complex features
- Architecture changes
- First-of-a-kind designs
- Changes with significant risk implications

V.3 Design Review Record

Design review records should include:

- What was reviewed (feature, design document, architecture)
- Who participated (names, roles)
- Date of review
- Issues identified
- Disposition of issues (who is responsible, when addressed)
- Conclusion (approved, approved with conditions, not approved)

> **483 PATTERN:** Design reviews conducted without quality involvement. Inspector sees engineering design reviews but no cross-functional participation. 820.30(e) requires individuals without direct responsibility for the design stage being reviewed, which typically means quality involvement.

Appendix W — Training and Competency Requirements

This appendix addresses training requirements for framework implementation.

W.1 Role-Based Training

Table W-1: Role-Based Training Requirements

Role	Training Topics	Competency Demonstration
DQE	ISO 14971, CCB process, verification standards	Classification exercises, audit scenarios
Product Owner	SRS requirements, traceability, regulatory constraints	Requirements writing exercises
Engineering Lead	Design control basics, evidence requirements	Sprint DHF packet review
PM	CCB facilitation, tracker maintenance	CCB simulation, tracker accuracy verification
RA	Submission impact assessment, regulatory updates	Scenario analysis

W.2 New Employee Onboarding

New team members require framework orientation before participating in sprints:

Day 1: Overview of design control requirements and why they exist

Day 2: Framework structure - CCB, sprint integration, evidence requirements

Day 3: Tool training - Jira configuration, tracker usage

Governance qualifier: Jira (or any ALM tool) can be the system of record for design control evidence if — and only if — it preserves contemporaneous evidence, full change history, review records, and traceability links that cannot be backdated or deleted. The tool is not the system; the discipline is the system. The tool must make the discipline visible and auditable.

Day 4: Role-specific training per table above

Day 5: Observation of CCB meeting and sprint activities

Week 2: Supervised participation in sprint

Week 3+: Independent participation with mentor check-ins

W.3 Ongoing Competency

Annual refresher training should cover:

- Framework updates and lessons learned
- Regulatory updates affecting the framework
- 483 patterns and how to avoid them
- Classification calibration exercises

483 PATTERN: Training records don't demonstrate competency. Records show employees attended training but no evidence they understood the material. Competency demonstration (exercises, tests, practical application) is required.

Appendix X — Audit Preparation Checklist

This appendix provides a comprehensive checklist for audit preparation.

X.1 Pre-Audit Documentation Review

□ All Sprint DHF packets assembled and indexed

□ Traceability reports current

□ CCB records complete with rationale

□ Risk file current version available

□ SRS current version available

□ Test protocols and results indexed

□ Training records current

□ CAPA records current with effectiveness evidence

□ Complaint log current

□ PMS reports current

X.2 Sample Selection Preparation

Auditors will select samples. Be prepared to demonstrate:

For any change:

□ Show CCB classification record

□ Show sprint where implemented

□ Show requirements linkage

□ Show risk file entries

□ Show verification evidence

□ Show approval records

□ Demonstrate timestamps interleave

For any complaint:

□ Show complaint record

□ Show disposition/investigation

□ Show trend analysis

□ If CAPA: show design change and effectiveness

X.3 Personnel Preparation

Identify who answers which questions. This is not administrative—it is a critical audit preparation step. Auditors are trained to notice hesitation, inconsistency, and deferred answers. When the wrong person attempts to explain a risk classification or verification decision, the explanation tends to be vague, secondhand, or contradicted by documentation. Pre-assigning Subject Matter Experts by topic area ensures that the person who actually made the decision is the person who explains it. Table X-1 provides the baseline mapping.

Table X-1: SME Quick Reference

Topic	Primary SME	Backup SME
Risk management	DQE	RA
Requirements	Product Owner	PM
Design decisions	Engineering Lead	Architect
Testing	Test Lead	DQE
CCB process	PM	DQE
Complaints/PMS	PMS Manager	DQE
CAPA	Quality Manager	DQE

SMEs should review recent activities in their area and be prepared to walk through specific examples. The most common audit failure mode in personnel preparation is not absence of knowledge—it is absence of a concrete example. When an auditor asks "how do you determine risk classification?" the DQE who answers with the general process but cannot walk through a specific recent item is less credible than one who says "here is item STORY-24-006, classified MEDIUM on March 3rd because it affects user workflow but not clinical decision-making—here is the CCB record." Have each SME identify two or three recent examples they can walk through end-to-end before the audit begins.

X.4 Common Auditor Questions

Be prepared for these frequently asked questions:

1. Walk me through how this change was controlled.
2. Who determines risk classification and how?
3. Show me evidence that the risk file was current during development.
4. How do complaints feed back into design?
5. What happens if an item fails verification?
6. Show me your traceability for this feature.
7. How do you ensure regression testing is adequate?
8. Who approves verification adequacy?

Appendix Y — Sprint Ceremonies in Regulated Environment

This appendix provides detailed guidance on adapting standard Agile ceremonies for regulated software development.

Y.1 Sprint Planning

Pre-Planning CCB Queue Review

Before sprint planning, ensure all candidate items have been classified at CCB. Items without Safety_Impact classification cannot enter the sprint.

PM reviews CCB queue 2-3 days before sprint planning:

- Items awaiting classification → Add to CCB agenda
- Items classified but missing required fields → Follow up with owners
- Items ready for sprint → Confirm in backlog

Sprint Planning Meeting Structure

Duration: 2-4 hours depending on sprint length and team size

Part 1: What can we deliver? (1-2 hours)

- Product Owner presents prioritized backlog
- For each item, review Safety_Impact classification
- DQE confirms rigor requirements based on classification
- Team discusses capacity and dependencies
- Sprint goal established

Part 2: How will we deliver it? (1-2 hours)

- Break items into tasks
- Identify DQE involvement points for HIGH/MEDIUM items
- Identify verification approach for each item
- Identify documentation deliverables (SRS updates, risk file updates)
- Commit to sprint scope

Output: Sprint backlog with all items classified, rigor requirements identified, and verification approach defined.

Planning Checklist

□ All items have Safety_Impact classification

□ HIGH items have RA consultation scheduled if needed

□ DQE availability confirmed for HIGH/MEDIUM items

□ Verification approach identified for each item

□ Documentation owners assigned (SRS, Risk, Test)

□ Sprint goal documented

Y.2 Daily Standup

Standup Participation

Standard participants: Engineering team, Product Owner, Scrum Master/PM

DQE participation:

- Required for sprints with HIGH items
- Recommended for sprints with MEDIUM items
- Optional for sprints with only LOW items

When DQE attends, they should speak to:

- Risk file update status
- Any risk concerns identified during reviews
- Verification readiness

Standup Questions for Regulated Environment

Standard questions plus:

- Is documentation keeping pace with implementation?
- Are there any risk implications from yesterday's work?
- Is verification planning on track?

Escalation Triggers

Immediately escalate if:

- Implementation reveals new hazards not in risk file
- Requirements need significant change after classification
- Verification approach proves inadequate
- External dependency threatens evidence completion

Y.3 Sprint Review

Review Structure

Duration: 1-2 hours

Part 1: Demonstration (45-60 min)

- Demonstrate completed features
- Product Owner confirms acceptance criteria met
- Stakeholders provide feedback

Part 2: Evidence Review (30-45 min)

- Review evidence completion status from tracker
- DQE presents verification summary for HIGH/MEDIUM items
- Identify any gaps requiring carryover

Output: Verified increment with complete evidence, or identified gaps for resolution.

Review Checklist

□ All committed items demonstrated or accounted for

□ Tracker shows current status for all items

□ HIGH items have DQE verification approval

□ MEDIUM items have DQE verification approval

□ Evidence links populated for all items

□ Carryover items identified with reason

Y.4 Sprint Retrospective

Retrospective Topics for Regulated Teams

Standard retrospective topics plus:

Evidence timing:

- Did documentation keep pace with implementation?
- Were there any last-minute documentation scrambles?
- How can we improve parallel execution?

CCB effectiveness:

- Were classifications appropriate?
- Did rigor levels match actual complexity?
- Were there any surprises during implementation?

DQE integration:

- Was DQE availability sufficient?
- Were risk discussions timely?
- Did verification approval happen smoothly?

Retrospective Metrics to Review

- Evidence completion rate (% of items with complete evidence at sprint end)
- Documentation lag (days between implementation and documentation)
- Verification cycle time (days from implementation to DQE approval)
- CCB classification accuracy (items requiring reclassification)

Appendix Z — Metrics and Measurement

This appendix provides guidance on measuring framework effectiveness and identifying improvement opportunities.

Z.1 Leading Indicators

Leading indicators predict future compliance status. Monitor these to catch problems early.

CCB Queue Age

Definition: Days items spend in CCB queue before classification

Target: <3 days average

Warning: >5 days average

Action if warning: Increase CCB frequency or streamline classification criteria

Evidence Completion Rate at Sprint End

Definition: % of completed items with all required evidence at sprint close

Target: >95%

Warning: <90%

Action if warning: Investigate parallel execution breakdown, DQE availability, documentation ownership

Documentation Lag

Definition: Average days between code commit and documentation update

Target: <2 days

Warning: >5 days

Action if warning: DQE not embedded in sprint, or documentation ownership unclear

Verification Cycle Time
Definition: Days from implementation complete to DQE approval

Target: <3 days

Warning: >5 days

Action if warning: DQE bottleneck, unclear verification criteria, or test environment issues

Z.2 Lagging Indicators

Lagging indicators measure past performance. Use these for trend analysis and improvement validation.

Release Gate Pass Rate
Definition: % of releases that pass Release Gate on first attempt

Target: >90%

Warning: <80%

Interpretation: Low pass rate indicates sprint-level execution problems—evidence not completing during sprints

Audit Finding Rate
Definition: Number of findings per audit related to design control

Target: 0 major findings, <3 minor findings

Warning: Any major finding

Interpretation: Audit findings reveal systemic issues the metrics above should have caught earlier

Retrofit Tax Measurement
Definition: Hours spent on documentation activities after sprint close

Target: <5% of sprint capacity

Warning: >15% of sprint capacity

Interpretation: This directly measures the problem the framework solves. If retrofit tax persists, the framework isn't being followed.

Z.3 Dashboard Design

The metrics defined in Sections Z.1 and Z.2 are most effective when consolidated into a single compliance dashboard reviewed at a consistent cadence. Table Z-1 provides the recommended dashboard structure. Each row corresponds to a metric from the leading or lagging indicator set, with columns for the current measured value, the target threshold, the directional trend since last measurement, and the current status classification. The Current Value column is populated during weekly leadership reviews using data pulled from the Design Control Tracker. Trend is assessed by comparing the current value to the prior period: improving (moving toward target), stable (unchanged), or declining (moving away from target). Status is color-coded: green (on target), yellow (warning—approaching the action threshold), or red (critical—action threshold exceeded). When any metric shows yellow or red, the leadership team should identify the root cause and assign corrective action before the next review cycle.

Table Z-1: Compliance Dashboard Metrics

Metric	Current Value	Target	Trend	Status	Action Threshold
CCB Queue Age	(days)	< 3 days	▲ ► ▼	● ● ●	> 5 days
Evidence Completion Rate	(%)	> 95%	▲ ► ▼	● ● ●	< 90%
Documentation Lag	(days)	< 2 days	▲ ► ▼	● ● ●	> 5 days
Verification Cycle Time	(days)	< 3 days	▲ ► ▼	● ● ●	> 5 days
Release Gate Pass Rate	(%)	> 90%	▲ ► ▼	● ● ●	< 80%

Legend: **Trend:** ▲ Improving ► Stable ▼ Declining **Status:** ● On Target ● Warning ● Critical

Enter current values during weekly leadership reviews. Color-code status based on current value vs. target and action threshold.

Review this dashboard weekly in leadership meetings. The value of the dashboard is not the snapshot—it is the trend. A single yellow metric is a data point. Three consecutive weeks of yellow in the same metric is a systemic issue requiring investigation. When a metric crosses from yellow to red, escalate to the responsible role immediately: CCB Queue

Age escalation goes to PM, Evidence Completion to DQE, Documentation Lag to PM and DQE jointly, Verification Cycle to the Test Lead, and Release Gate Pass Rate to the Program Manager. The five metrics shown in Table Z-1 are the minimum recommended set. Organizations may add metrics specific to their context—for example, SOUP vulnerability response time for teams with significant third-party dependencies, or CAPA cycle time for organizations under active remediation.

Z.4 Continuous Improvement

Use metrics to drive continuous improvement:

Monthly: Review metric trends. Identify any declining patterns.

Quarterly: Deep dive on metrics outside target. Identify root causes. Define improvement actions.

Annually: Comprehensive framework review. Are the targets still appropriate? Should processes evolve?

Appendix AA — Organizational Change Management

Implementing this framework requires organizational change. This appendix addresses common change management challenges.

AA.1 Common Resistance Patterns

Engineering Resistance

Pattern: 'This is just more bureaucracy slowing us down.'

Response: The framework actually reduces total effort by eliminating the retrofit tax. Show the data: teams spend 20-40% of release effort on post-development documentation. The framework eliminates this.

Key message: You're already doing the thinking. This framework just captures it as you go instead of reconstructing it later.

QA Resistance

Pattern: 'We can't operate at sprint speed. We need time to review properly.'

Response: The DQE role is designed to operate at sprint speed. This requires embedding DQE in the sprint, not reviewing after. If DQE reviews during the sprint, the review happens when context is fresh—faster and more accurate than reviewing afterward.

Key message: Same rigor, different timing. You're not reviewing less—you're reviewing earlier.

Leadership Resistance

Pattern: 'We've been doing it this way for years. Why change now?'

Response: Regulatory expectations are evolving. AAMI TIR45 now explicitly addresses Agile. FDA investigators are increasingly sophisticated about software development. The retrofit approach that worked 10 years ago creates audit risk today.

Key message: The question isn't whether to change—it's whether to lead the change or be forced into it by a regulatory finding.

AA.2 Stakeholder Mapping

Table AA-1: Stakeholder Engagement Matrix

Stakeholder	Primary Concern	Key Message	Engagement Approach
Engineering	Speed, autonomy	Less rework, faster releases	Demonstrate time savings
QA/DQE	Rigor, thoroughness	Same rigor, better timing	Pilot sprint with metrics
Product	Features, roadmap	Faster releases, fewer delays	Show release velocity improvement
RA	Compliance, submissions	Better evidence, stronger position	Audit scenario walkthrough
Leadership	Cost, risk	Reduced audit risk, lower cost	ROI analysis and risk reduction

AA.3 Pilot Approach

Don't roll out to the entire organization at once. Use a pilot approach:

Phase 1 (Weeks 1-4): Single team pilot

- Select one team willing to try the framework
- Implement full framework for that team
- Measure metrics before and after
- Document lessons learned

Phase 2 (Weeks 5-8): Expand to 2-3 teams

- Apply lessons from pilot
- Refine processes based on feedback

- Build internal expertise

Phase 3 (Weeks 9-12): Organization-wide rollout

- Formalize procedures
- Train all teams
- Establish ongoing support

AA.4 Success Criteria for Pilot

Define success before starting the pilot:

Quantitative:

- Evidence completion rate >90% at sprint end
- Documentation lag <3 days average
- No increase in sprint velocity loss
- Retrofit tax reduction >50%

Qualitative:

- Team reports process is sustainable
- DQE reports adequate involvement
- No 'documentation scramble' at release
- Mock inspection demonstrates audit-ready evidence

Appendix BB — CAPA Integration with Design Control

Corrective and Preventive Action (CAPA) is a separate quality system that intersects with design control. This appendix clarifies the integration.

BB.1 When CAPA Triggers Design Change

Not all CAPAs require design changes. CAPA may result in:

- Training updates (no design change)
- Process improvements (no design change)
- Design modifications (design change required)
- Product recalls (design change likely required)

When CAPA determines a design change is needed:

1. CAPA owner creates Intake signal: 'CAPA-XXXX requires design change'
2. Signal enters CCB queue like any other change
3. CCB classifies based on patient safety impact (typically HIGH for CAPAs)
4. Design change proceeds through normal sprint process
5. Evidence links back to CAPA
6. CAPA effectiveness criteria include design change verification

BB.2 CAPA Evidence Requirements

When design changes result from CAPA, the design history must include:

- CAPA reference (which CAPA drove this change)
- Root cause analysis summary (why the change is appropriate)
- Verification that change addresses root cause
- Effectiveness monitoring plan
- Effectiveness evidence when monitoring complete

BB.3 CAPA Effectiveness and Design Control

CAPA effectiveness verification often depends on design control evidence:

Immediate effectiveness: Did the design change pass verification?

Short-term effectiveness: Did complaints of this type decrease?

Long-term effectiveness: Has the issue recurred?

The design control system provides evidence for immediate effectiveness. The PMS system provides evidence for short-term and long-term effectiveness.

> **REGULATORY ANCHOR:** 21 CFR 820.100 requires that CAPA be effective. For design-related CAPAs, effectiveness requires demonstrating that the design change addresses the root cause and prevents recurrence. The design control evidence (verification, validation) demonstrates immediate effectiveness; PMS monitoring demonstrates sustained effectiveness.

> **483 PATTERN:** CAPA closed without design control evidence. CAPA record says 'design change implemented' but no reference to design history, no verification evidence, no traceability. Inspector cannot confirm the change actually occurred or was properly controlled.

Appendix CC — Supplier and Third-Party Component Control

When design includes third-party components, supplier design control extends your design control obligations. This appendix addresses supplier integration.

CC.1 Supplier Classification

Classify suppliers based on component criticality:

Table CC-1: Supplier Classification Matrix

Supplier Type	Criteria	Control Level
Critical	Component affects clinical functionality	Full design control visibility
Major	Component affects user workflow	Change notification required
Minor	Infrastructure/commodity	Standard vendor management

CC.2 Critical Supplier Requirements

For critical suppliers (components affecting clinical functionality):

Design control visibility:

- Access to supplier's design history for the component
- Notification of design changes before release
- Verification evidence for changes affecting your device
- Risk management integration (supplier hazards in your risk file)

Change control:

- Supplier changes enter your Intake
- CCB classifies impact on your device
- Your verification includes supplier component verification

Quality agreement:

- Document requirements in supplier quality agreement
- Audit rights for design control records
- Notification requirements for changes

CC.3 Off-the-Shelf Software Components

For SOUP (Software of Unknown Provenance) and OTS (Off-the-Shelf) components:

Risk assessment required:

- What hazards could this component contribute to?
- What is the component's failure mode?
- What is the impact on your device if the component fails?

Verification approach:

- Functional testing at integration level
- Published anomaly list review
- Version control and change monitoring

> **REGULATORY ANCHOR**: IEC 62304:2006+A1:2015 Clause 8 addresses software configuration management including SOUP. See Appendix AAA for comprehensive IEC 62304 governance including clause-by-class applicability and SOUP requirements (Clauses 5.3.3, 5.3.4, 8.1.2). You must identify SOUP items, evaluate their risk contribution, and verify their performance in your system context.

CC.4 Open Source Software

Open source components require additional considerations:

License compliance:

- Review license terms for each component
- Ensure compliance with distribution requirements
- Document license obligations

Security monitoring:

- Monitor CVE databases for vulnerabilities
- Establish update process for security patches
- Document security assessment

Support and maintenance:

- Assess project health (active maintenance?)
- Plan for component becoming unmaintained
- Consider fork/internalization strategy

Appendix DD — Integrating Legacy Systems

Many organizations have existing products developed under different processes. This appendix addresses integrating legacy systems into the framework.

DD.1 Assessment Approach

Before integrating a legacy product, assess current state:

Documentation state:

- Does a DHF exist? How complete is it?
- Are requirements documented? Current?
- Does a risk file exist? When was it last updated?
- Is traceability established?

Process state:

- How have changes been controlled historically?
- Are there undocumented changes?
- What verification exists?

DD.2 Gap Remediation

Based on assessment, determine gap remediation approach:

Option 1: Full remediation

- Reconstruct DHF to current standards
- Update risk file completely
- Establish full traceability
- Cost: High. Benefit: Complete compliance.

Option 2: Point-forward remediation

- Accept current state as baseline
- Apply framework to all future changes
- Gradually improve documentation through normal maintenance
- Cost: Lower. Benefit: Progressive improvement.

Option 3: Risk-based remediation

- Prioritize remediation based on risk
- Full remediation for high-risk areas
- Point-forward for lower-risk areas
- Cost: Medium. Benefit: Risk-proportionate.

DD.3 Baseline Establishment

Regardless of remediation approach, establish a baseline:

- Document current software version as baseline
- Document known requirements (even if incomplete)
- Document known risks (even if incomplete)
- Document known verification status
- All future changes go through CCB and full framework

The baseline acknowledges: 'We know what we know and what we don't know. Going forward, everything is controlled.'

> **483 PATTERN:** Legacy product changes not controlled. Organization claims product was 'developed before current procedures' but continues making changes without design control. Legacy status doesn't exempt ongoing changes from current requirements.

Appendix EE — Scaling for Different Organization Sizes

This appendix addresses how the framework scales for different organization sizes.

EE.1 Small Teams (5-15 people)

In small organizations, individuals often hold multiple roles.

Role Combinations

Table EE-1: Role Combination Guidelines

Combined Role	Responsibilities	Caution
Eng Lead + Product Owner	Technical decisions + requirements	Avoid self-approving requirements
DQE + RA	Sprint classification + regulatory strategy	Ensure both functions get attention
PM + Scrum Master	Sprint mechanics + CCB facilitation	Natural combination

Process Simplifications

- CCB can be informal (DQE decision with PM recording) for LOW items
- Sprint DHF packet can be simplified folder structure
- Tracker can be Excel-based rather than Jira dashboard

What Cannot Be Simplified

- Classification must still happen before sprint
- Evidence must still be created during sprint
- Verification approval must still be independent
- Traceability must still exist

EE.2 Medium Organizations (15-50 people)

Medium organizations typically have dedicated roles but may share across products.

Shared Resources

- One DQE may support multiple product teams
- RA typically shared across organization

- CCB may cover multiple products

Coordination Challenges

- DQE availability across multiple sprints
- CCB scheduling for multiple product streams
- Consistent classification across products

Solutions

- Stagger sprint schedules so DQE isn't overloaded
- Use product-specific CCB queues, shared CCB meeting
- Maintain classification reference library across products

EE.3 Large Organizations (50+ people)

Large organizations have dedicated roles but face coordination complexity.

Common Patterns

- Dedicated DQE per product team
- DQE community of practice for consistency
- Product-specific CCBs with enterprise oversight
- Shared RA function with product liaisons

Governance Considerations

- Enterprise design control procedure (umbrella)
- Product-specific work instructions
- Cross-product consistency audits
- Classification calibration across DQEs

Tool Requirements

- Enterprise Jira configuration
- Cross-product reporting dashboards
- Shared document control system
- Audit trail across products

Appendix FF — Detailed Jira Configuration Guide

This appendix provides step-by-step Jira configuration instructions for implementing the framework.

FF.1 Custom Field Setup

Safety_Impact Field

Type: Select List (single choice)

Options: HIGH, MEDIUM, LOW

Required: Yes (for all issue types)

Screen: All screens

Default: None (force explicit selection)

Configuration steps:

1. Jira Settings → Issues → Custom Fields → Create Custom Field

2. Select 'Select List (single choice)'

3. Name: 'Safety_Impact'

4. Add options: HIGH, MEDIUM, LOW (in that order)

5. Associate with relevant screens

6. Make required via workflow validator

CCB_Reference Field

Type: URL or Text Field

Required: Yes (when Safety_Impact is set)

Screen: All screens

Purpose: Link to CCB meeting record or minutes

SRS_Link Field

Type: URL

Required: No (optional for LOW items)

Screen: All screens

Purpose: Link to relevant SRS section in document control system

Risk_Link Field
Type: URL

Required: No (optional for LOW items)

Screen: All screens

Purpose: Link to relevant hazard entry in risk file

Verification_Status Field
Type: Select List (single choice)

Options: NOT_STARTED, IN_PROGRESS, PASSED, FAILED, APPROVED, N/A

Required: Yes for MEDIUM/HIGH items at Done transition

Screen: All screens

FF.2 Workflow Configuration

Standard Workflow States

- Backlog: Item created, awaiting prioritization
- CCB_Queue: Item prioritized, awaiting classification
- Classified: Item classified by CCB, in sprint backlog
- In_Progress: Item in active sprint work
- In_Review: Implementation complete, verification in progress
- Done: All work and verification complete
- Blocked: Item cannot proceed due to impediment (returns to CCB_Queue for re-evaluation)

Key Transitions and Validators

CCB_Queue → Classified:

Validator: Safety_Impact is not empty

Validator: CCB_Reference is not empty

Post-function: Set transition timestamp

Classified → In_Progress:

Validator: Safety_Impact is set

Condition: Sprint is assigned

In_Review Decision (Blocked?):

If No (not blocked) → Done:

Validator: If Safety_Impact = HIGH or MEDIUM, then Verification_Status must = APPROVED

Validator: Evidence links populated based on classification

Post-function: Set completion timestamp

If Yes (blocked) → Blocked:

Condition: Item has impediment (external dependency, failing tests, or other blocker)

Blocked → CCB_Queue:

Condition: Impediment resolved

Action: Returns to CCB for re-evaluation (scope or risk may have changed)

Workflow Diagram

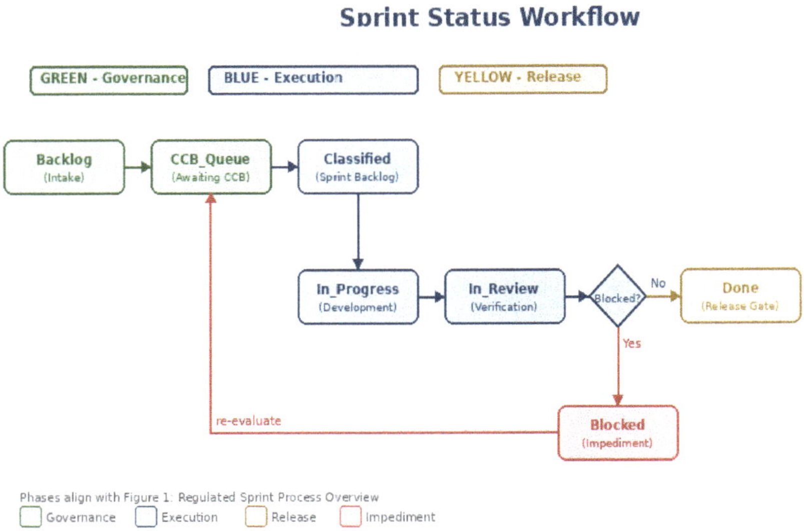

Figure FF-1: Sprint Status Workflow — Jira States with Phase Color Coding

This diagram aligns with Figure 1: Regulated Sprint Process Overview, using the same GREEN (Governance), BLUE (Execution), and YELLOW (Release) phase color coding.

FF.3 Dashboard Configuration

Sprint Compliance Dashboard

Gadget 1: Pie chart - Safety_Impact distribution

Filter: Sprint = current AND status in (Classified, In_Progress, In_Review)

Gadget 2: Two-dimensional filter - Status vs Safety_Impact

Shows count of items by status and classification

Gadget 3: Filter results - Blocked items

Filter: Sprint = current AND status = Blocked

Gadget 4: Filter results - Missing evidence

Filter: Sprint = current AND status = Done AND (SRS_Link is EMPTY OR Risk_Link is EMPTY)

Release Readiness Dashboard
Gadget 1: Statistics - Total items, Ready count, Blocked count

Gadget 2: Filter results - HIGH items not APPROVED

Filter: Fix_Version = target AND Safety_Impact = HIGH AND Verification_Status != APPROVED

Gadget 3: Filter results - MEDIUM items not APPROVED

Filter: Fix_Version = target AND Safety_Impact = MEDIUM AND Verification_Status != APPROVED

Gadget 4: Pie chart - Verification status distribution

FF.4 Automation Rules

Rule 1: CCB Queue Notification
Trigger: Issue moved to CCB_Queue

Action: Send email to DQE and PM

Action: Add comment 'Awaiting CCB classification'

Rule 2: HIGH Risk Alert
Trigger: Safety_Impact set to HIGH

Action: Send email to RA

Action: Add label 'HIGH_RISK'

Rule 3: Evidence Reminder
Trigger: Issue in In_Review for >2 days AND Verification_Status = NOT_STARTED

Action: Send email to assignee and DQE

Action: Add comment 'Evidence reminder: Verification pending'

Appendix GG — Extended Mock Inspection Scenarios

This appendix provides additional mock inspection scenarios for audit preparation.

GG.1 Design Input Scenarios

MOCK INSPECTION Q&A: Show me the design inputs for this feature.

Weak answer: "The product owner wrote up the requirements in this document."

Strong answer: "Here's the complete design input chain. User needs were captured during our formative usability research—here's the usability report documenting those needs. Each user need traces to one or more SRS requirements. Here's the traceability showing User Need UN-023 links to SRS requirements 4.3.1 through 4.3.5. The SRS was approved before verification completed, consistent with AAMI TIR45 guidance for iterative development."

MOCK INSPECTION Q&A: How do you ensure design inputs are appropriate for the intended use?

Weak answer: "Our product team understands the clinical workflow."

Strong answer: "Design input appropriateness is verified through multiple mechanisms. First, user needs are derived from clinical stakeholder input and formative usability studies with actual clinicians. Second, our RA reviews intended use alignment for any requirements affecting clinical claims. Third, our DQE reviews requirements for risk implications before CCB classification. Here's an example where DQE flagged a requirement as potentially creating a new hazard—the requirement was revised before implementation."

MOCK INSPECTION Q&A: This requirement seems vague. How do you verify it?

Weak answer: "Our test team interprets the requirement and writes appropriate tests."

Strong answer: "You're right that requirement could be more specific. During our sprint, the acceptance criteria were refined to be testable. Here's the Jira ticket showing the original requirement and the refined acceptance criteria added during sprint planning. The test cases directly reference these acceptance criteria. We've identified requirements quality as an improvement area and are implementing acceptance criteria reviews at sprint planning."

GG.2 Risk Management Scenarios

MOCK INSPECTION Q&A: Walk me through your hazard identification process.

Weak answer: "We brainstorm potential hazards during design reviews."

Strong answer: "Hazard identification is systematic. For each clinical function, we apply our hazard identification checklist which covers categories like wrong patient, wrong result, wrong timing, missing information, and unintended actions. Here's the checklist. For feature X, we applied this checklist and identified these hazards. The process is documented in our risk management plan section 5.2, and the hazard analysis shows the systematic application."

MOCK INSPECTION Q&A: How do you determine probability estimates?

Weak answer: "We use our engineering judgment based on experience."

Strong answer: "Probability estimation combines multiple inputs. P1 estimates consider frequency of the hazardous situation—how often is the user exposed to this hazard? We use usability data when available, field data for similar features, and engineering analysis of failure modes. P2 estimates consider the likelihood of harm given exposure. Here's an example: for hazard HA-0034, our P1 estimate is based on the usability study showing 2 of 15 users encountered this situation, and our P2 estimate is based on clinical severity if they proceed without the information."

MOCK INSPECTION Q&A: I see training listed as your primary risk control. Did you consider design alternatives?

Weak answer: "Training is effective for this type of issue."

Strong answer: "We evaluated risk control options per the ISO 14971 hierarchy. Here's the risk control evaluation record for this hazard. We first considered inherent safety by design—could we eliminate the hazard? The analysis shows why elimination wasn't feasible for this hazard. We then considered protective measures—here's the design alternative we evaluated and why it wasn't implemented. Only after documenting why higher-priority options weren't feasible did we select training as the control. The training is reinforced by on-screen reminders, which is a protective measure layered with the training."

GG.3 Verification Scenarios

MOCK INSPECTION Q&A: Show me the verification evidence for this requirement.

Weak answer: "Let me find the test results."

Strong answer: "Here's the complete verification chain. Requirement SRS-4.2.3 links to test case TC-4.2.3-001. The test case was executed on March 15th—here's the execution record showing the test environment, tester, inputs, expected results, actual results, and pass determination. The test case also verifies risk control RC-0023. DQE approved verification adequacy on March 16th—here's the approval record."

MOCK INSPECTION Q&A: How do you ensure your test environment matches production?

Weak answer: "We try to keep them similar."

Strong answer: "Test environment configuration is controlled. Here's our test environment specification document. For each release, we document the test environment configuration and compare against production configuration. Differences are evaluated for impact on verification validity. Here's the environment comparison for Release 4.2—the only difference was database size, which we addressed with specific performance tests using production-scale data."

MOCK INSPECTION Q&A: This test failed initially. What happened?

Weak answer: "We fixed the bug and reran the test."

Strong answer: "Correct. Here's the complete record. The initial test execution on March 10th failed—actual result didn't match expected result. The failure was investigated and determined to be an implementation defect. Bug ticket BUG-4589 was created, classified at CCB as MEDIUM, and fixed in the same sprint. The test was re-executed on March 14th and passed. Both execution records are retained, showing the complete verification history including the failure and resolution."

GG.4 Change Control Scenarios

MOCK INSPECTION Q&A: How do you handle urgent changes that can't wait for CCB?

Weak answer: "We have an expedited process that skips CCB."

Strong answer: "We have an expedited process that accelerates CCB, not skips it. For urgent changes, we convene an emergency CCB within hours—DQE and PM at minimum. Classification still happens before work begins. Here's an example: Critical security vulnerability identified Monday morning, emergency CCB at 10 AM, classified as HIGH due to potential patient data exposure, fix implemented and deployed by Wednesday. All evidence exists—just on an accelerated timeline."

MOCK INSPECTION Q&A: I see this change was classified LOW but it affects the display format. Why?

Weak answer: "Display changes are always LOW."

Strong answer: "Actually, that's not our policy—display changes are classified based on clinical impact. Let me look at this specific classification. Here's the CCB record... The rationale states this display change affects footer formatting only, with no impact on clinical information display. The clinical data display area was unchanged. If a display change affected clinical information presentation, it would be MEDIUM or HIGH. This classification appears appropriate based on the scope of the change."

GG.5 Post-Market Scenarios

MOCK INSPECTION Q&A: How do complaints inform your risk analysis?

Weak answer: "We track complaints separately from risk."

Strong answer: "Complaints directly update our risk analysis. When a complaint indicates a hazardous situation occurred in the field, our PMS Manager reviews the risk file entry for that hazard. If the complaint reveals a hazardous situation we hadn't identified, we add it. If the complaint provides data to refine our P1 estimate, we update it. Here's an example: these three complaints about timeout behavior informed our P1 estimate for HA-0048—field frequency is higher than our initial estimate, so we updated the risk file and prioritized a fix."

MOCK INSPECTION Q&A: Show me how a CAPA resulted in a design change.

Weak answer: "Let me find the CAPA record."

Strong answer: "Here's CAPA-2024-0023. Root cause analysis identified a design deficiency—the timeout handling didn't account for large datasets. The corrective action was a design change, which entered our Intake on February 15th. CCB classified it as MEDIUM on February 17th. The fix was implemented in Sprint 50. Here's the design change record with traceability to the CAPA. Effectiveness was monitored for 90 days—zero recurrence at sites with the updated software. CAPA was closed on May 20th with effectiveness confirmed."

Appendix HH — Document Templates

This appendix provides structured templates for key framework documents. Use these as starting points—adapt field names and sections to match your organization's quality system.

HH.1 CCB Meeting Minutes Template

Field	Content	Notes
Meeting Date	[DATE]	Format: YYYY-MM-DD
Facilitator	[PM Name]	PM owns CCB scheduling
Attendees	[Names and Roles]	Minimum: DQE, PM, Engineering Lead
RA Present	[Yes/No]	Required when HIGH items on agenda
Classification Rationale	Hazard: [device characteristic] / Hazardous Situation: [exposure/use error] / Harm: [clinical outcome]	Required for MEDIUM and HIGH; recommended for all
P1 Basis	[Data/source supporting exposure probability estimate — usability data, literature, field data]	Required for HIGH; recommended for MEDIUM
P2 Basis	[Clinical rationale for harm probability given exposure; risk control effectiveness evidence]	Required for HIGH; recommended for MEDIUM

Per-Item Classification Record:

Field	Content	Regulatory Basis
Item ID	[Jira ticket number]	Traceability requirement (820.30(i))
Item Title	[Brief description of change]	
Requester	[Name]	
Description	[What the change does — 1-2 sentences]	
Classification	[HIGH / MEDIUM / LOW]	ISO 14971 Clause 6.2
Rationale	[Why this classification — reference universal question]	Must not be blank
DQE	[DQE name who made determination]	Single point of accountability
RA Consultation	[Required / Not Required / Completed]	Required for HIGH items
Action Items	[Any follow-up needed before sprint entry]	

Meeting Close:

Field	Content
Total Items Classified	[Count]
HIGH Items	[Count and IDs]
MEDIUM Items	[Count]
LOW Items	[Count]
Action Items	[List with owners and due dates]
Next CCB Date	[DATE]

HH.2 Sprint DHF Packet Cover Sheet Template

Field	Content
Sprint Number	[Sprint ID]
Sprint Dates	[Start Date] to [End Date]
Product	[Product Name]
Software Version	[Version at sprint close]

Packet Contents Checklist:

Document	Included	Version/Reference	Notes
Sprint Summary	☐		Items completed, classifications
CCB Records	☐	[CCB-YYYY-NNNN]	One record per classified item
Requirements (SRS)	☐	[vX.X]	Sprint-end version
Design Documentation (SDS)	☐	[vX.X]	If architecture changes
Risk Documentation	☐	[vX.X]	Sprint-end version of risk file
Verification Evidence	☐		Test protocol and results
Approval Records	☐		DQE, RA (if HIGH items)
Traceability Report	☐		Generated from Jira links

Risk Summary:

Classification	Count	Item IDs
HIGH	[n]	[List IDs]
MEDIUM	[n]	[List IDs]
LOW	[n]	[List IDs]

Approvals:

Role	Name	Signature	Date	Required When
DQE	[Name]		[Date]	Always
RA	[Name]		[Date]	HIGH items in sprint

HH.3 Release Gate Checklist Template

Field	Content
Release Version	[Version]
Gate Date	[Date]
Included Sprints	[Sprint N through Sprint M]
Program Manager	[Name]

Pre-Conditions:

Condition	Status	Evidence Reference	Notes
All Sprint DHF Packets complete	☐ Pass / ☐ Fail		
Tracker shows required status for all items	☐ Pass / ☐ Fail		
No BLOCKED items remain	☐ Pass / ☐ Fail		
Version baseline established	☐ Pass / ☐ Fail		

Evidence Review:

Check	Status	Exceptions
All items have Safety_Impact classification	☐ Pass / ☐ Fail	
All HIGH items: DQE + RA approval	☐ Pass / ☐ Fail	
All MEDIUM items: DQE approval	☐ Pass / ☐ Fail	
All evidence links populated	☐ Pass / ☐ Fail	
Traceability report generated and reviewed	☐ Pass / ☐ Fail	
HIGH items: benefit-risk conclusion documented (most likely vs worst plausible harm; P1 and P2 addressed)	☐ Pass / ☐ Fail / ☐N/A (no HIGH items)	
Post-market signals: all signals mapped to causal chain IDs in risk file (no parallel taxonomy)	☐ Pass / ☐ Fail / ☐N/A	

Special Considerations:

Item	Status	Details
Unresolved complaints affecting release	☐ None / ☐ See details	
Open CAPAs affecting release	☐ None / ☐ See details	
Known limitations	☐ None / ☐ See details	
Validation required	☐ Complete / ☐ N/A	[Report ref if applicable]

Release Decision:

Decision	Details
☐ APPROVED for release	
☐ NOT APPROVED	[Reason and required actions]

Release Approvals:

Role	Name	Signature	Date	Required When
DQE	[Name]		[Date]	Always
RA	[Name]		[Date]	HIGH items in release
Program Manager	[Name]		[Date]	Always — owns release decision

NOTE: These templates provide the minimum required fields. Organizations should add fields as needed for their quality system. All templates should be controlled documents within your QMS.

Appendix II — Frequently Asked Questions

This appendix addresses common questions about the framework.

II.1 General Framework Questions

Q: Do we need to implement everything at once?
A: No. Start with the core elements: CCB classification, DQE at sprint cadence, and Definition of Done with three sign-offs. Add refinements (dashboards, automation, detailed procedures) over time. The pilot approach (Appendix AA) provides a phased implementation path.

Q: How does this work with SAFe or other scaled Agile frameworks?
A: The framework is agnostic to Agile scaling approach. In SAFe, the CCB can operate at team level with PI-level oversight. The core principles (classify before sprint, evidence during sprint, approve before release) apply regardless of scaling framework.

Q: What if our QA department resists the DQE model?
A: The DQE model doesn't replace QA—it adds sprint-cadence quality involvement. Traditional QA activities (audit, CAPA, supplier quality) continue. DQE is a specific accountability for sprint-level risk classification and verification approval. One person might hold both QA and DQE accountabilities, or they might be separate roles.

II.2 CCB Questions

Q: How long should CCB meetings take?
A: A well-functioning CCB processes most items in 2-3 minutes each. A 10-item CCB should take 30-45 minutes. If CCB routinely takes longer, classification criteria may be unclear or items aren't coming in prepared.

Q: What if DQE and Engineering disagree on classification?
A: DQE makes the classification decision—it's not a vote. However, DQE should consider Engineering's input about technical implications. If there's persistent disagreement, it suggests classification criteria need clarification. Document the disagreement and rationale in the CCB record.

Q: Can we classify items in advance without a meeting?
A: For clearly LOW items similar to previously classified items, DQE can classify asynchronously with CCB ratification at the next meeting. This speeds throughput while maintaining oversight. HIGH items should always be discussed live.

II.3 Evidence Questions

Q: What counts as 'evidence created during sprint'?
A: Evidence created during sprint means documentation that reflects understanding at the time of development. If the risk file update shows the DQE thinking about hazards while the design was being finalized, that's evidence during sprint. If the risk file update is a retroactive justification written after implementation was complete, that's not.

Q: How do we handle evidence for legacy items?
A: For items developed before the framework was implemented, establish a baseline acknowledging current documentation state. Going forward, all changes to those items flow through the framework. You don't need to retroactively create evidence for decisions that were made years ago, but you do need to control current changes.

Q: Can we use automated testing as verification evidence?
A: Yes, automated test results are valid verification evidence. The key requirements: test cases must be traceable to requirements, test execution must be documented (when, what version, what environment), and results must be retained. Automated testing can provide better evidence than manual testing because execution is consistent and complete.

II.4 Tool Questions

Q: We don't use Jira. Can we use other tools?
A: Yes. The principles apply to any work tracking system. You need: custom fields for classification and traceability, workflow enforcement to prevent bypass, and audit trail of changes. Azure DevOps, Monday,

Asana, or even paper-based systems can work if they meet these requirements.

Q: Our document control system doesn't integrate with Jira. How do we handle traceability?
A: Use URL links. The SRS_Link and Risk_Link fields contain URLs to documents in your document control system. The traceability is maintained through these links. It's not as automated as a fully integrated system, but it works. The key is that links must be maintained and verified.

II.5 Regulatory Questions

Q: Will FDA accept this approach?
A: This framework implements design controls per 21 CFR 820.30 and risk management per ISO 14971. It's consistent with AAMI TIR45 guidance on iterative development. FDA doesn't prescribe specific processes—they evaluate whether your system produces compliant evidence. This framework is designed to produce that evidence.

Q: How does this work for EU MDR?
A: The framework produces evidence that can populate the EU MDR Technical Documentation structure (Annex II). See Appendix I for detailed mapping. The main differences are format requirements and some additional MDR-specific content (PRRC, UDI, EUDAMED), but the underlying design control evidence is the same.

Q: What about international registrations beyond FDA and EU?
A: Most international regulatory requirements for design control are based on ISO 13485 and ISO 14971, which this framework implements. Country-specific requirements typically add to rather than replace these standards. The framework provides a foundation that can be extended for specific markets.

Appendix JJ — Clinical Decision Support (CDS) Considerations

Clinical Decision Support software has specific regulatory and design control considerations. This appendix addresses CDS-specific requirements.

JJ.1 CDS Definition and Classification

FDA's 2022 CDS guidance establishes criteria for determining whether CDS is a medical device. The four criteria for non-device CDS:

1. Displays or conveys patient-specific information from the EHR or other sources

2. Displays or conveys knowledge or clinical practice guideline recommendations

3. Allows the healthcare provider to independently review the basis for the recommendation

4. Healthcare provider does not primarily rely on the CDS recommendation to make clinical decisions

If any criterion is not met, the CDS is likely a medical device requiring design control.

CDS Classification Matrix

Table JJ-1: Clinical Decision Support Classification

CDS Type	Typical Classification	Design Control Required
Simple alert based on threshold	Meets all 4 criteria	No (non-device)
Drug interaction checker with references	Meets all 4 criteria	No (non-device)
Risk score with algorithm transparency	Likely meets criteria	Maybe (analyze each criterion)

CDS Type	Typical Classification	Design Control Required
AI-derived recommendation	Likely device	Yes (Class II typical)
Treatment recommendation	Likely device	Yes (Class II typical)
Diagnostic recommendation	Device	Yes (Class II typical)

JJ.2 CDS Hazard Categories

CDS introduces specific hazard categories that should be systematically considered:

Information Quality Hazards

- Incorrect calculation or scoring
- Outdated clinical guidelines or reference data
- Data quality issues from source systems
- Missing or incomplete patient data

Presentation Hazards

- Alert fatigue leading to ignored warnings
- Information presented in confusing format
- Critical information not prominent
- Timing delays in displaying information

Integration Hazards

- EHR data mapping errors
- Timestamp or patient matching issues
- Synchronization failures between systems
- Version incompatibilities with source systems

JJ.3 CDS-Specific Verification

CDS verification should include:

Clinical accuracy verification:

- Compare outputs against known correct results
- Verify calculations match published algorithms
- Validate against clinical expert review
- Test boundary conditions and edge cases

Guideline currency verification:

- Document clinical guideline sources
- Verify guidelines are current
- Establish process for guideline updates
- Track guideline changes as design inputs

Integration verification:

- Verify correct data mapping from EHR
- Test with realistic clinical data
- Verify correct patient identification
- Test error handling for bad or missing data

> **REGULATORY ANCHOR:** FDA's 2022 CDS guidance clarifies that CDS meeting all four criteria is not a device. However, the guidance explicitly states that safety and effectiveness remain the manufacturer's responsibility regardless of device status. Good design practices are still appropriate even for non-device CDS.

Appendix KK — AI/ML Software as Medical Device

AI/ML-enabled SaMD introduces unique design control challenges. This appendix provides detailed guidance for AI/ML integration.

KK.1 AI/ML Lifecycle in Design Control

Traditional software has a linear development cycle: requirements → design → implementation → verification. AI/ML introduces additional cycles:

Training cycle:

- Training data collection and curation
- Model architecture selection
- Training execution
- Model evaluation
- Model selection

This training cycle has design control implications:

- Training data requirements are design inputs
- Model architecture is design output
- Training execution is implementation
- Model evaluation is verification

KK.2 Training Data as Design Input

Training data characteristics become design inputs that must be documented:

Data source specification:

- Where does training data come from?
- What patient populations are represented?
- What clinical settings are represented?
- What time periods are represented?

Data quality requirements:

- What label quality is required?
- How are labels verified?
- What completeness requirements apply?
- How are errors and inconsistencies handled?

Bias considerations:

- Are all relevant populations represented?
- Are there potential sources of bias in labeling?
- How is bias evaluated and mitigated?

> **REGULATORY ANCHOR:** FDA's AI/ML Action Plan and Good Machine Learning Practice (GMLP) guiding principles emphasize representative training data. Training data characteristics are design inputs that affect device performance.

KK.3 Model Performance as Design Output

Model performance becomes a design output that must be verified:

Performance metrics:

- Sensitivity and specificity (for classifiers)
- Area under ROC curve
- Precision and recall
- Performance across subgroups

Performance acceptance criteria:

- What performance is acceptable?
- How is performance evaluated?
- How does performance compare to predicate or clinical benchmark?

KK.4 Predetermined Change Control Plan (PCCP)

FDA's PCCP framework allows pre-specifying types of modifications that can be made without new submission. For AI/ML:

PCCP may cover:

- Model retraining with similar data
- Performance improvement within specified bounds
- Feature additions that don't change clinical intent
- Updates to address performance drift

PCCP must specify:

- Description of modifications planned
- How modifications will be implemented
- How modifications will be assessed
- Labeling changes that may result

Changes within PCCP:

- Still require design control
- Still require CCB classification (typically MEDIUM)
- Still require verification per PCCP specifications
- Don't require new regulatory submission

Changes outside PCCP:

- Require HIGH classification
- Require RA consultation
- Likely require new regulatory submission

KK.5 Continuous Learning Systems

Some AI/ML systems learn continuously from production data. This creates unique design control challenges:

When does learning require design control?

- Any change to model parameters = design change
- Continuous learning = continuous design change
- Must be within PCCP boundaries or pre-specified algorithm

Monitoring requirements:

• Performance monitoring to detect drift

• Input distribution monitoring to detect data shift

• Outcome monitoring if available

• Automatic alerts for performance changes

Lock/unlock model patterns:

• Locked model: Frozen after initial training (simplest to control)

• Periodic update: Model retrained on schedule (batch control)

• Continuous learning: Model updates continuously (requires automatic controls)

> **KEY INSIGHT:** Continuous learning doesn't eliminate design control—it requires automated design control. If the model updates automatically, the controls must be automatic too: performance bounds, automatic rollback, alert thresholds.

Causal discipline reminder: Environment contributes to hazardous situation; hazard remains the device characteristic (including data handling, logic, and availability design choices). Input distribution shift and model drift describe environmental or operational conditions that surface a hazardous situation — the underlying hazard is the device's design characteristic (e.g., insufficient generalization, absence of out-of-distribution detection, or inadequate performance bounds).

Appendix LL — Interoperability and Integration

SaMD increasingly operates as part of larger healthcare IT ecosystems. This appendix addresses interoperability-specific design control considerations.

LL.1 Integration Hazard Categories

Integration introduces specific hazard categories:

Data mapping hazards:

- Fields mapped incorrectly between systems
- Unit conversions performed incorrectly
- Coded values translated incorrectly
- Optional fields handled inconsistently

Patient identification hazards:

- Wrong patient data associated with wrong patient
- Patient merge/split scenarios not handled
- Duplicate records not detected

Timing hazards:

- Stale data used for clinical decisions
- Timestamp synchronization issues
- Order-of-arrival assumptions violated

Availability hazards:

- Dependent system unavailable
- Network interruptions during data exchange
- Partial data received

LL.2 Interface Requirements

Interface specifications are design inputs that must be documented:

For each interface, document:

- Connected system identification
- Data exchanged (in and out)

- Protocol and format (HL7, FHIR, proprietary)
- Frequency and trigger conditions
- Error handling approach
- Timeout and retry behavior

LL.3 Integration Verification

Integration verification should cover:

Connectivity verification:

- Connection establishment under normal conditions
- Connection handling under adverse conditions
- Reconnection after interruption

Data exchange verification:

- Correct data mapping for all fields
- Correct handling of optional/missing data
- Correct handling of boundary values
- Correct handling of coded values and terminologies

Error handling verification:

- Graceful degradation when partner system unavailable
- User notification of integration failures
- Data integrity when errors occur
- Recovery after error conditions

> **REGULATORY ANCHOR:** FDA's Interoperability and Related Capabilities guidance emphasizes that integration risks must be part of risk management. Interface-related hazards and controls should appear in the risk file.

Appendix MM — Mobile and Distributed Software

SaMD deployed on mobile devices or distributed across multiple components has additional considerations. This appendix addresses these scenarios.

MM.1 Mobile Platform Considerations

Platform Dependencies

Mobile SaMD depends on the mobile platform (iOS, Android). This creates design control considerations:

Platform version requirements:

- Minimum supported OS version
- Maximum supported OS version (if any)
- Impact of platform updates on device function
- Process for validating new platform versions

Platform capability dependencies:

- Hardware features used (camera, GPS, sensors)
- Platform services used (notifications, background processing)
- Behavior when capabilities unavailable

App Store Considerations

Distribution through app stores introduces:

- Review and approval dependencies
- Timing uncertainty for releases
- Version control across user population
- Forced updates vs. user-controlled updates

MM.2 Client-Server Architecture

When SaMD spans client and server components:

Version compatibility:

- Which client versions work with which server versions?
- How are incompatible combinations handled?

- What happens during rolling deployments?

Responsibility boundaries:

- Which hazards are addressed by client?
- Which hazards are addressed by server?
- What are the interface hazards between them?

Verification scope:

- Client-only testing
- Server-only testing
- End-to-end integration testing
- Version matrix testing

MM.3 Offline Operation

If SaMD can operate offline:

Offline capability requirements:

- What functions are available offline?
- What data is cached locally?
- How long can offline operation continue?
- What clinical limitations apply offline?

Synchronization requirements:

- How is data synchronized when connectivity returns?
- How are conflicts resolved?
- What happens to actions taken offline?

Risk considerations:

- Stale data risks during offline operation
- Data loss risks during synchronization
- User awareness of offline status

Appendix NN — Documentation Management Best Practices

Effective documentation management is essential for the framework. This appendix provides detailed guidance.

NN.1 Document vs. Record Distinction

Documents are controlled by version; records are controlled by retention. Understanding this distinction is crucial:

Table NN-1: Document vs Record Characteristics

Characteristic	Document	Record
Examples	SRS, Risk File, SDS, Procedures	Test results, CCB minutes, Approvals
Change approach	New version replaces old	Original retained, new record added
Version control	Version number incremented	Not versioned (timestamped)
Approval	Re-approved when changed	Approved once at creation
Retention	Current version + change history	Original retained for retention period

NN.2 Document Control Requirements

For controlled documents (SRS, Risk File, SDS, procedures):

Identification:

- Unique document number
- Version number
- Title
- Effective date
- Author/owner

Review and approval:

- Defined review process
- Defined approval authority
- Documented approval (signature or electronic equivalent)

Change control:

- Changes require approval
- Change history maintained
- Reason for change documented

Distribution:

- Controlled distribution to authorized users
- Obsolete versions removed from circulation
- Users acknowledge receipt of updates

NN.3 Electronic Documentation

Electronic document control systems must meet specific requirements:

Access control:

- User authentication
- Role-based permissions
- Edit rights limited to authorized users

Audit trail:

- All changes logged
- Change author identified
- Change timestamp recorded
- Change reason captured (if required)

Version control:

- Version history maintained
- Prior versions accessible

• Current version clearly identified

Electronic signatures (if used):

• 21 CFR Part 11 compliance

• Signature bound to document version

• Signer identification verified

• Signature meaning documented

NN.4 Documentation Review Practices

Effective documentation requires regular review:

Periodic review:

• Annual review of procedures

• Risk file review at least annually

• Update documents when underlying information changes

Triggered review:

• Design changes trigger requirement/risk review

• Regulatory changes trigger procedure review

• Audit findings trigger relevant document review

• Field signals trigger risk file review

> **483 PATTERN:** Documents show last review date years in the past. Inspector asks 'when was this risk file last reviewed?' Answer is 3+ years ago. Risk file should be reviewed at least annually and updated when design changes occur.

Appendix OO — Inspection Preparation Deep Dive

This appendix provides comprehensive guidance for FDA inspection preparation.

OO.1 Pre-Inspection Preparation Timeline

6 Months Before Expected Inspection

- Review recent 483s issued to similar companies
- Conduct internal audit of design control system
- Identify and prioritize gaps
- Begin gap remediation
- Update procedures as needed

3 Months Before

- Complete gap remediation
- Conduct mock inspection
- Train personnel on inspection protocols
- Prepare document index and location guide
- Identify subject matter experts for each area

1 Month Before

- Refresh training for front-line personnel
- Verify all documentation is current
- Prepare war room or document staging area
- Review recent changes that may be sampled
- Prepare standard responses for common questions

1 Week Before

- Final document location verification
- SME availability confirmation
- Communication plan finalized
- Logistics prepared (conference room, printing, etc.)

OO.2 During the Inspection

Opening Meeting

- Introduce key personnel and roles
- Provide company overview relevant to inspection scope
- Offer facility tour if appropriate
- Clarify logistics and availability
- Don't volunteer information beyond questions asked

Document Requests

- Confirm understanding of what's requested
- Provide exactly what's requested (not more)
- Document everything provided
- If document doesn't exist, say so clearly
- If time needed to locate, provide timeline

Questions and Interviews

- Answer questions directly
- If you don't know, say so and offer to find out
- Don't speculate or guess
- Don't answer questions beyond what's asked
- If question is unclear, ask for clarification

Issues Identified

- Don't argue or become defensive
- Acknowledge observations factually
- Provide context if relevant
- Don't commit to specific corrective actions on the spot
- Document observations for follow-up

Daily Wrap-up

- Internal debrief each day
- Discuss any issues identified

- Prepare for next day's focus areas
- Begin researching any pending questions

OO.3 Post-Inspection

Closeout Meeting

- Receive preliminary observations
- Clarify any misunderstandings
- Don't negotiate findings
- Thank inspector for their time

483 Response (if issued)

- Respond within 15 business days
- For each observation:
  - Acknowledge the observation
  - Describe root cause analysis
  - Describe corrective actions taken or planned
  - Provide timeline for completion
  - Include evidence of completed actions

Lessons Learned

- Document inspection observations
- Identify systemic issues
- Update procedures as needed
- Share lessons across organization
- Update inspection preparation for next time

Appendix PP — Statistical Methods in Design Verification

This appendix provides guidance on statistical methods applicable to SaMD design verification.

PP.1 When Statistics Apply

Statistical methods apply when verification involves:

- Performance measurement against specifications (sensitivity, specificity)
- Reliability estimation (failure rates, mean time between failures)
- Sample-based testing (when exhaustive testing isn't feasible)
- Comparison to predicates or benchmarks

Statistical methods don't apply when:

- Testing is deterministic (same input always produces same output)
- Complete enumeration is feasible (all test cases can be executed)
- Requirements are binary (feature present or absent)

PP.2 Sample Size Determination

When verification involves sampling, sample size must be justified:

For performance estimation (e.g., sensitivity):

- Specify desired confidence level (typically 95%)
- Specify acceptable margin of error
- Calculate sample size from statistical formula
- Document sample size justification in test protocol

For reliability demonstration:

- Specify reliability claim (e.g., 99% reliability)
- Specify confidence level
- Calculate sample size for zero-failure or accept/reject plans
- Document demonstration protocol

PP.3 Performance Metrics for Classifiers

AI/ML classifiers require specific verification metrics:

Table PP-1: AI/ML Verification Metrics

Metric	Definition	When Important
Sensitivity	True positive / (True positive + False negative)	When missing positives is critical
Specificity	True negative / (True negative + False positive)	When false alarms are costly
PPV	True positive / (True positive + False positive)	When positive results drive action
NPV	True negative / (True negative + False negative)	When negative results rule out condition
AUC	Area under ROC curve	Overall discriminative ability

Document acceptance criteria for each relevant metric. Verify against independent test set not used in training.

PP.4 Equivalence and Non-Inferiority

When comparing to predicates or benchmarks:

Equivalence testing:

• Define equivalence margin (clinically acceptable difference)

• Demonstrate difference is within margin (both directions)

• Requires larger sample sizes than superiority testing

Non-inferiority testing:

• Define non-inferiority margin

• Demonstrate new device is no worse than reference by more than margin

• Common for substantial equivalence claims

> **REGULATORY ANCHOR:** FDA's guidance on clinical performance assessment for diagnostic devices addresses statistical requirements for performance claims. Acceptance criteria should reflect clinical significance, not just statistical significance.

Appendix QQ — Design Transfer for SaMD

Design transfer for SaMD differs from hardware transfer. This appendix addresses SaMD-specific considerations.

QQ.1 What Design Transfer Means for Software

For hardware, design transfer means moving from engineering to manufacturing. For software, the concept is different:

Traditional software transfer:

- Development environment to production environment
- Development team to operations/support team
- One-time deployment vs. managed release process

SaMD transfer considerations:

- Build and deployment processes must be validated
- Production environment must match verified configuration
- Support processes must be established
- Monitoring and surveillance must be active

QQ.2 Build Process Validation

The software build process is the 'manufacturing process' for software:

Build reproducibility:

- Same source code produces same binary (deterministic builds)
- Build environment is controlled and documented
- Dependencies are version-locked
- Build process is automated and repeatable

Build verification:

- Automated tests run as part of build
- Build failures prevent release
- Build artifacts are versioned and traceable

QQ.3 Deployment Validation

Deployment processes must ensure correct software reaches production:

Deployment verification:

- Deployed version matches intended version
- Configuration matches verified configuration
- Post-deployment verification confirms correct operation

Rollback capability:

- Ability to revert to previous version if issues found
- Rollback process tested and documented
- Rollback decision authority defined

QQ.4 Support Transition

Transfer includes transitioning from development team to support:

Knowledge transfer:

- Architecture documentation current
- Known issues documented
- Troubleshooting guides prepared
- Support training completed

Support processes:

- Incident response procedures established
- Escalation paths defined
- On-call responsibilities assigned
- Monitoring and alerting active

> **REGULATORY ANCHOR:** 21 CFR 820.30(h) requires procedures for design transfer. For software, this means validated build processes, controlled deployment, and established support processes.

Appendix RR — Extended 483 Pattern Analysis

This appendix provides deeper analysis of 483 patterns with root cause investigation and systematic prevention.

RR.1 Pattern Category: Temporal Evidence Failures

> **483 PATTERN:** All design documentation timestamps fall on release date. Inspector recognizes that documents were created retroactively rather than during development.

Root cause analysis:

• Documentation treated as release activity rather than development activity

• No DQE embedded in sprints

• No parallel execution discipline

• Cultural norm of 'documentation comes last'

Systematic prevention:

• Embed DQE in sprint ceremonies

• Require documentation updates during sprint, not after

• Track documentation lag as leading indicator

• Include documentation currency in Definition of Done

RR.2 Pattern Category: Classification Failures

> **483 PATTERN:** Changes affecting clinical functionality classified as LOW or not classified at all. No systematic safety assessment before implementation.

Root cause analysis:

• Classification based on requester's perception, not objective analysis

• No consistent classification criteria

• CCB bypassed or rubber-stamped

• Insufficient technical analysis at classification

Systematic prevention:

• Standard classification question applied consistently

• DQE independence from implementation team

• Classification reference library for consistency

• Periodic classification audits

RR.3 Pattern Category: Traceability Failures

> **483 PATTERN:** Requirements exist. Tests exist. But no documented link between them. Cannot demonstrate that requirements were verified.

Root cause analysis:

• Requirements and tests in separate systems without links

• No ownership of traceability maintenance

• Traceability treated as compliance activity rather than engineering practice

• No validation of traceability completeness

Systematic prevention:

• Links created at same time as test case creation

• Jira links as single source of traceability

• Traceability report as release gate requirement

• Incomplete traceability blocks release

RR.4 Pattern Category: Risk Management Failures

> **483 PATTERN:** Risk file exists but doesn't reflect current design. Hazards from years ago remain; recent design changes not reflected.

Root cause analysis:

• Risk file created for initial release, never updated

• Design changes not systematically evaluated for risk impact

• No ownership of risk file maintenance

• Risk management perceived as one-time activity

Systematic prevention:

• Risk file update required for all changes (even if 'no impact')

• DQE owns risk file currency

• Risk_Link field in Jira proves contemporaneous update

• Risk file review included in release gate

RR.5 Pattern Category: Change Control Failures

483 PATTERN: Design history shows large gaps between changes. Evidence suggests changes occurred but weren't controlled. 'Undocumented changes' discovered during inspection.

Root cause analysis:

• 'Engineering only' exceptions allowed

• Urgent changes bypassed process

• Process perceived as too slow, so avoided

• No technical enforcement of change control

Systematic prevention:

• Jira workflow enforcement (no bypass possible)

• No 'engineering only' category

• Expedited process for urgent changes (still controlled)

• Regular audits comparing deployed code to controlled changes

Appendix SS — Case Study: Framework Implementation

This appendix presents a fictional but realistic case study of framework implementation.

SS.1 Starting Situation

Company: MedTech Software Inc.

Product: Diagnostic support software (Class II SaMD)

Team: 25 engineers, 5 QA, 2 RA

Development: Scrum, 2-week sprints

Pain points:

- 30% of release effort spent on documentation 'catch-up'
- Releases consistently delayed 2-3 weeks for compliance
- Engineering and QA in constant friction
- Recent warning letter citing design control deficiencies

SS.2 Assessment Phase (2 weeks)

Team assessed current state:

- CCB existed but only for HIGH-level changes (most changes bypassed)
- Risk file updated at release time (typically 6-8 weeks after design decisions)
- Traceability created manually at release (error-prone, time-consuming)
- No DQE role—QA operated at audit cadence, not sprint cadence

Gap analysis identified:

- No systematic classification of all changes
- Documentation lag averaged 4 weeks behind implementation
- Traceability incomplete for 20% of requirements
- No visibility into compliance status during sprints

SS.3 Implementation Phase (12 weeks)

Weeks 1-4: Foundation

- Created DQE role (reassigned from QA)

- Added 5 custom fields to Jira
- Configured workflow to require Safety_Impact before In Progress
- Scheduled weekly CCB
- Trained team on framework concepts

Weeks 5-8: Pilot Sprint

- Selected one product team for pilot
- Ran 4 sprints with full framework
- DQE attended all standups
- Classified all items at CCB before sprint entry
- Generated Sprint DHF packets at sprint close

Pilot metrics:

- Documentation lag reduced from 4 weeks to 3 days
- Evidence completion at sprint end: 92%
- No increase in sprint velocity loss
- Retrofit time reduced by 75%

Weeks 9-12: Rollout

- Applied lessons learned from pilot
- Extended to all product teams
- Refined CCB cadence (twice weekly for high-volume teams)
- Created compliance dashboard
- Documented procedures

SS.4 Results After 6 Months

Quantitative results:

- Documentation lag: <2 days average (down from 4 weeks)
- Evidence completion at sprint end: >95%
- Retrofit time: <5% of release effort (down from 30%)
- Release delays: Eliminated
- Audit findings: Zero design control findings in subsequent audit

Qualitative results:

- Engineering reports less friction with compliance
- QA reports better visibility into design decisions
- Releases feel 'ready' at release time—no scramble
- Confidence in audit readiness significantly increased

SS.5 Lessons Learned

What worked:

- Starting with pilot team built expertise before broad rollout
- DQE role essential—sprint cadence quality requires dedicated role
- Jira workflow enforcement prevented bypass
- Metrics visibility drove improvement

What was challenging:

- Initial resistance from engineers ('more process')
- DQE availability constraint across multiple teams
- Classification consistency required ongoing calibration
- Legacy products needed baseline establishment

Recommendations for others:

- Don't try to implement everything at once
- Measure before and after to demonstrate value
- Invest in the DQE role—it's the linchpin
- Enforce through tools, not just policy

Appendix TT — FDA QMSR and ISO 13485:2016 Integration

In February 2024, FDA finalized the Quality Management System Regulation (QMSR), which harmonizes 21 CFR Part 820 with ISO 13485:2016. This is a significant regulatory development that affects all medical device manufacturers, including SaMD developers. This appendix provides detailed guidance on QMSR requirements and how this framework aligns with both the traditional Part 820 and the new ISO 13485-based QMSR.

TT.1 The QMSR Transition

FDA's QMSR represents a fundamental shift in U.S. medical device quality system regulation:

What changed:

• Part 820 now incorporates ISO 13485:2016 by reference

• FDA retains certain 'additions' beyond ISO 13485 requirements

• Compliance deadline: February 2, 2026 (2-year transition)

• Organizations must comply with QMSR, not choose between Part 820 and ISO 13485

• FDA Compliance Program CP 7382.850 (effective February 2, 2026) provides a primary inspection framework

• Inspections now organized around six QMS Areas with “risk” at center

Why it matters for SaMD:

• Global harmonization reduces burden for international manufacturers

• Risk-based thinking is emphasized throughout ISO 13485:2016

• Process approach aligns well with iterative software development

• Design control requirements remain substantively similar

> **REGULATORY ANCHOR:** FDA QMSR (21 CFR Part 820, effective February 2026) incorporates ISO 13485:2016 by reference.

Manufacturers must comply with ISO 13485:2016 plus FDA-specific additions covering management responsibility, CAPA, MDR reporting, and other requirements.

TT.2 ISO 13485:2016 Design Control Requirements

ISO 13485:2016 Clause 7.3 addresses design and development. The requirements align closely with the original 21 CFR 820.30:

Table TT-1: ISO 13485:2016 Clause 7.3 Mapping

ISO 13485:2016 Clause	Requirement	Framework Implementation
7.3.1	Design and development planning	Sprint planning, CCB, tracker
7.3.2	Design and development inputs	SRS, user needs, risk inputs
7.3.3	Design and development outputs	SDS, implemented software, verification evidence
7.3.4	Design and development review	Sprint reviews, Definition of Done, Release Gate
7.3.5	Design and development verification	Test execution, DQE approval
7.3.6	Design and development validation	Validation sprint, usability validation
7.3.7	Control of design changes	CCB classification, change control records
7.3.8	Design and development transfer	Build validation, deployment procedures
7.3.9	Design and development files	Sprint DHF packets, traceability

TT.3 Key ISO 13485:2016 Additions

ISO 13485:2016 introduced several clarifications that affect SaMD development:

Risk-Based Approach

ISO 13485:2016 Clause 0.2 states that 'the term risk is used, the application of the term within the scope of this International Standard pertains to safety or performance requirements of the medical device.' This aligns with the framework's emphasis on patient safety as the classification criterion at CCB.

Process Approach

ISO 13485:2016 Clause 0.3 describes the process approach: 'Any activity that receives input and converts it to output can be considered as a process. Often the output from one process directly forms the input to the next process.'

The framework's loop (Intake → CCB → Sprint → Release → Field → PMS → Intake) is exactly this process approach. Each stage produces outputs that become inputs to the next stage.

Documented Information

ISO 13485:2016 Clause 4.2.4 requires 'Records shall remain legible, readily identifiable and retrievable.' The Sprint DHF packet approach ensures records are organized, traceable, and retrievable.

TT.4 FDA QMSR Additions

FDA retains certain requirements beyond ISO 13485:2016. These 'additions' include:

Table TT-2: FDA Additions Beyond ISO 13485

Topic	FDA Addition	Framework Compliance
Management Representative	Must be member of management	Program Manager or designated leader
Quality Policy	Must include commitment to meet requirements	Standard quality policy language
CAPA	Enhanced requirements for investigation and effectiveness	PMS → CAPA → Design Change loop
Complaint Handling	Must comply with 21 CFR 820.198	Complaint triage per PMS process
MDR Reporting	Must comply with 21 CFR Part 803	PMS Manager responsibility
Corrections and Removals	Must comply with 21 CFR Part 806	PMS/RA coordination
UDI	Must comply with 21 CFR Part 830	RA responsibility at release

TT.5 Framework Mapping to ISO 13485:2016

This section maps framework activities to specific ISO 13485:2016 clauses:

Clause 4: Quality Management System

Table TT-3: ISO 13485 Clause 4 Mapping

Subclause	Requirement	Framework Element
4.1	General requirements	Overall framework structure
4.1.3	Outsourced processes	Supplier control (Appendix CC)
4.2.1	Documentation requirements	SRS, SDS, Risk File, procedures
4.2.3	Medical device file	DHF = Sprint DHF packets
4.2.4	Control of documents	Document control system
4.2.5	Control of records	Record retention per procedures

Clause 7.3: Design and Development

Table TT-4: ISO 13485 Clause 7.3 Mapping

Subclause	Requirement	Framework Element
7.3.1	Planning	Sprint planning, CCB classification
7.3.2	Inputs	SRS, user needs, risk inputs at CCB
7.3.3	Outputs	Verified software, test evidence, updated risk file

Subclause	Requirement	Framework Element
7.3.4	Review	Definition of Done, sprint review
7.3.5	Verification	Test execution with DQE approval
7.3.6	Validation	Validation sprint, usability testing
7.3.7	Change control	CCB classification, traceability
7.3.8	Transfer	Build validation, deployment
7.3.9	Files	Sprint DHF packets assembled per release

Clause 7.5.6: Production Process Validation
For software, 'production' is the build process. ISO 13485:2016 Clause 7.5.6 requires validation of processes where resulting output cannot be verified by subsequent monitoring or measurement.

Framework compliance: Build process validation (Appendix QQ) ensures reproducible, verified builds.

TT.6 Transition Considerations

Organizations transitioning to QMSR should consider:

Gap analysis:

- Compare current QMS to ISO 13485:2016 requirements
- Identify documentation that needs updating
- Identify processes that need modification
- Plan training for personnel on ISO 13485:2016 terminology

Documentation updates:

- Quality Manual may need restructuring
- Procedures may need ISO 13485:2016 clause references
- Forms and templates may need updating
- Training records need updating

Timeline:

- QMSR effective date: February 2, 2026
- Organizations should complete transition before this date
- Third-party certifications to ISO 13485:2016 can support transition

TT.7 CP 7382.850 Framework Alignment

FDA's Compliance Program CP 7382.850, effective February 2, 2026, provides the primary inspection framework for medical device manufacturers under QMSR. The program organizes inspections around six QMS Areas with patients/users and risk at the center. This framework maps directly to each QMS Area:

Change Control QMS Area: Framework's CCB classification, risk gate, and change control records (Chapters 9, 11)

Design and Development QMS Area: Regulated sprint, parallel execution, Sprint DHF packets (Chapters 8, 10, Appendix M)

Management Oversight QMS Area: Role definitions, RACI matrix, quality culture (Chapter 7, Appendix C)

Measurement, Analysis, and Improvement QMS Area: PMS process, complaint handling, CAPA integration (Chapters 13-14, Appendix BB)

Outsourcing and Purchasing QMS Area: Supplier control, SOUP management (Appendix CC)

Production and Service Provision QMS Area: Build validation, traceability, release gates (Chapter 15, Appendix QQ)

CP 7382.850 directs investigators to review risk management documentation across inspection areas, and notes that evaluation of one requirement may necessitate the evaluation of other requirements in different areas of the QMS. This is precisely how the framework operates—the regulated sprint loop (Intake → CCB → Sprint → Release → Field → PMS → Intake) demonstrates linked processes with risk at the center.

> **REGULATORY ANCHOR:** CP 7382.850 frames inspections around questions like: "Show me how a design change moved through your system." The framework's traceability from CCB classification through Sprint DHF packet to release gate directly answers this question with timestamps and evidence.

> **KEY INSIGHT:** The framework described in this book already aligns with ISO 13485:2016 design control requirements. Organizations following this framework should have minimal gaps when transitioning to QMSR.

Appendix UU — ISO 13485:2016 Clause-by-Clause Mapping

This appendix provides a detailed mapping of ISO 13485:2016 requirements to framework elements.

UU.1 Clause 4: Quality Management System

4.1 General Requirements

ISO 13485:2016 requires the organization to establish, document, implement, maintain, and continually maintain the effectiveness of a QMS.

Framework compliance:

- The regulated sprint process is the core QMS process for design control
- Procedures documented per Appendix HH templates
- Metrics (Appendix Z) demonstrate effectiveness maintenance
- Continuous improvement through retrospectives and CAPA

4.1.3 Outsourced Processes

ISO 13485:2016 requires control of outsourced processes affecting product conformity.

Framework compliance:

- Supplier control per Appendix CC
- Third-party component management
- Contract development considerations

4.2.3 Medical Device File

ISO 13485:2016 requires 'For each medical device type or medical device family, the organization shall establish and maintain one or more files either containing or referencing documents generated to demonstrate conformity to the requirements of this International Standard and compliance with applicable regulatory requirements.'

Framework compliance:

- Sprint DHF packets compile per-sprint evidence
- Release DHF assembles sprint packets
- Traceability reports demonstrate conformity
- Medical device file = DHF + regulatory documentation

UU.2 Clause 5: Management Responsibility

5.1 Management Commitment

ISO 13485:2016 requires top management commitment to QMS development, implementation, and maintenance.

Framework compliance:

- Program Manager role represents management commitment
- Release Gate requires management approval
- Resource allocation for DQE role demonstrates commitment

5.6 Management Review

ISO 13485:2016 requires management review at planned intervals.

Framework compliance:

- Quarterly or annual management reviews per procedures
- Review includes: metrics, CAPA status, audit findings, complaints
- Outputs include resource decisions, improvement actions

UU.3 Clause 6: Resource Management

6.2 Human Resources

ISO 13485:2016 requires personnel to be competent based on appropriate education, training, skills, and experience.

Framework compliance:

- Role-based training (Appendix W)
- Competency demonstration requirements
- Ongoing training for regulatory updates

6.3 Infrastructure

ISO 13485:2016 requires appropriate infrastructure including software.

Framework compliance:

- Jira configuration as infrastructure
- Document control system
- Build and test environments
- Tool validation (Appendix N)

UU.4 Clause 7: Product Realization

7.1 Planning of Product Realization

ISO 13485:2016 requires planning for product realization consistent with QMS requirements.

Framework compliance:

- Sprint planning addresses product realization planning
- CCB classification determines rigor requirements
- Quality objectives defined per release

7.2 Customer-Related Processes

ISO 13485:2016 requires determination and review of requirements related to the product.

Framework compliance:

- User needs capture and documentation
- Requirements review at CCB and sprint planning

• Customer feedback through PMS

7.3 Design and Development

See TT.2 and TT.5 for detailed 7.3 mapping.

7.4 Purchasing

ISO 13485:2016 requires control of purchasing processes.

Framework compliance:

• Supplier qualification per Appendix CC

• Incoming inspection/verification for components

• Supplier quality agreements

7.5 Production and Service Provision

For SaMD, 'production' is the build process.

Framework compliance:

• Build process control and validation

• Deployment procedures

• Servicing through maintenance releases

UU.5 Clause 8: Measurement, Analysis, and Improvement

8.2.1 Feedback

ISO 13485:2016 requires feedback processes as early warning of quality problems.

Framework compliance:

• PMS process captures feedback

• Complaint handling per procedures

• Field signals enter Intake

8.2.2 Complaint Handling

ISO 13485:2016 requires documented procedures for complaint handling.

Framework compliance:

- Complaint triage by PMS Manager
- Complaint → CAPA → Design Change pathway
- Regulatory reporting per FDA requirements

8.2.3 Reporting to Regulatory Authorities

ISO 13485:2016 requires reporting per applicable regulatory requirements.

Framework compliance:

- MDR reporting per 21 CFR Part 803 (FDA addition)
- RA responsibility for regulatory reporting
- PMS Manager identifies reportable events

8.3 Control of Nonconforming Product

ISO 13485:2016 requires control of nonconforming product.

Framework compliance:

- Software nonconformances = bugs/defects
- Bugs enter Intake and flow through CCB
- Release Gate prevents release of nonconforming product

8.4 Analysis of Data

ISO 13485:2016 requires analysis of data to demonstrate QMS suitability and effectiveness.

Framework compliance:

- Metrics per Appendix Z

- Trend analysis by PMS Manager
- Management review inputs

8.5 Improvement

ISO 13485:2016 requires identification and implementation of necessary changes.

Framework compliance:

- CAPA process per Appendix BB
- Retrospectives identify improvement opportunities
- Continuous improvement through metrics monitoring

Appendix VV — Updated Regulatory Reference: QMSR

This appendix supplements Appendix D with specific QMSR and ISO 13485:2016 references.

VV.1 FDA QMSR (21 CFR Part 820)

The Quality Management System Regulation, effective February 2, 2026, incorporates ISO 13485:2016 by reference with FDA-specific additions.

Key sections:

- 820.1 — Scope (incorporates ISO 13485:2016)
- 820.3 — Definitions (additional FDA definitions)
- 820.5 — Quality system (refers to ISO 13485:2016 Clause 4)
- 820.20 — Management responsibility (FDA additions to Clause 5)
- 820.25 — Personnel (refers to Clause 6.2)
- 820.30 — Design controls (refers to Clause 7.3) — RETAINED with additions
- 820.50 — Purchasing controls (refers to Clause 7.4)
- 820.70 — Production and process controls (refers to Clause 7.5)
- 820.90 — Nonconforming product (refers to Clause 8.3)
- 820.100 — CAPA (FDA additions to Clause 8.5)
- 820.198 — Complaint files (FDA-specific requirement)

VV.2 ISO 13485:2016 Key Clauses

ISO 13485:2016 'Medical devices — Quality management systems — Requirements for regulatory purposes'

Design-related clauses:

- 7.3.1 — Design and development planning
- 7.3.2 — Design and development inputs
- 7.3.3 — Design and development outputs
- 7.3.4 — Design and development review
- 7.3.5 — Design and development verification
- 7.3.6 — Design and development validation

• 7.3.7 — Control of design and development changes

• 7.3.8 — Design and development transfer

• 7.3.9 — Control of design and development files

• 7.3.10 — Design and development records

VV.3 Practical Implications

For organizations using this framework:

No major changes required:

• The framework already implements ISO 13485:2016 design control requirements

• Sprint DHF packets satisfy Clause 7.3.9 (design and development files)

• CCB satisfies Clause 7.3.7 (change control)

• Definition of Done satisfies Clause 7.3.4 (design review)

Documentation updates may be needed:

• Update procedure references from '21 CFR 820.30' to 'ISO 13485:2016 Clause 7.3 and 21 CFR 820.30'

• Consider ISO 13485:2016 terminology in procedures

• Ensure quality manual addresses ISO 13485:2016 structure

> **REGULATORY ANCHOR:** FDA QMSR aligns U.S. requirements with global ISO 13485:2016 standard. This framework implements both, providing global compliance foundation for SaMD manufacturers.

Appendix WW — IEC 62366-1 Usability Engineering Integration

IEC 62366-1:2015/AMD1:2020 'Medical devices — Part 1: Application of usability engineering to medical devices' is the international standard for usability engineering of medical devices. This appendix provides detailed guidance on integrating IEC 62366-1 requirements with the framework.

WW.1 IEC 62366-1 Overview

IEC 62366-1 establishes a usability engineering process that runs parallel to and integrates with design control and risk management. The standard focuses on use-related hazards—hazards that arise from the interaction between users and the device.

Key concepts:

• Use specification: Defines intended users, use environments, and user interface

• User interface: All means by which the user interacts with the device

• Use error: User action or lack of action that differs from what was expected

• Use-related hazard: Hazard that arises from the user interface

• Formative evaluation: Testing during design to improve the user interface

• Summative evaluation: Testing at end of design to verify user interface adequacy

> **REGULATORY ANCHOR:** IEC 62366-1:2015/AMD1:2020 Clause 5 requires a usability engineering process that identifies use-related hazards, implements risk controls in the user interface, and verifies the effectiveness of those controls through user testing.

WW.2 Usability Engineering File

IEC 62366-1 requires a usability engineering file (UEF) containing evidence of usability engineering activities. The UEF is distinct from but cross-references the DHF and risk management file.

UEF contents per IEC 62366-1:

- Use specification
- User interface specification
- Known or foreseeable hazards and hazardous situations related to usability
- Hazard-related use scenarios
- User interface evaluation plan
- Summative evaluation results
- Use-related risk analysis and residual risk evaluation

Framework integration:

- UEF is maintained as a controlled document alongside risk file
- Use specification feeds into SRS as user-related requirements
- Usability hazards appear in risk file with 'USE-' prefix
- Formative evaluation findings enter Intake as design inputs
- Summative evaluation is part of validation sprint

WW.3 Integration with ISO 14971 Causal Chain

IEC 62366-1 explicitly links to ISO 14971. The key integration point is that user behavior appears in the hazardous situation, not the hazard itself.

Correct causal chain example:

- Hazard (device characteristic): Alert displayed in small font
- Hazardous situation (includes user): Clinician fails to notice critical alert during high-workload period

• Harm: Patient receives delayed treatment

The risk control targets the hazard (device characteristic), not the user:

• Risk control: Increase alert prominence with larger font, color change, and audio notification

• Verification: Usability testing confirms users notice alert in high-workload scenarios

> **KEY INSIGHT:** IEC 62366-1 reinforces the principle that hazards are device characteristics. 'User error' is NEVER a hazard—it appears in the hazardous situation. Risk controls modify the device to prevent or reduce the occurrence of use errors.

WW.4 Formative Evaluation in Sprints

Formative evaluation is iterative testing during design to improve the user interface. This aligns naturally with sprint-based development.

Formative evaluation activities:

- Expert reviews (heuristic evaluation)
- Cognitive walkthroughs
- User testing with representative users
- Think-aloud protocols
- Usability inspections

Framework integration:

- Formative activities can occur within sprints
- Findings enter Intake as design inputs
- CCB classifies usability findings (typically MEDIUM)
- Fixes flow through normal sprint process
- No formal protocol required for formative testing

Table WW-1: Formative Usability Activities

Formative Activity	When	Output	Framework Entry Point
Heuristic evaluation	Early design	Usability issues list	Intake → CCB
Cognitive walkthrough	During sprint	Task completion issues	Sprint backlog
User testing (informal)	Mid-sprint or end	User feedback, errors observed	Intake → CCB
Expert review	Before release	Design recommendations	Intake → CCB

WW.5 Summative Evaluation

Summative evaluation is formal testing at the end of design to verify the user interface supports safe and effective use. This is a validation activity.

IEC 62366-1 Clause 5.9 requirements:

- Must test with representative users
- Must test in simulated or actual use environment
- Must address all hazard-related use scenarios
- Must follow documented protocol
- Results must demonstrate user interface adequacy

Framework integration:

- Summative evaluation occurs in validation sprint (Chapter 15)
- Code freeze during summative testing
- Protocol developed and approved before testing
- Results documented in usability engineering file
- Findings that require design changes → new sprint → re-validation

> **483 PATTERN:** Summative evaluation conducted without representative users. Organization tested with internal staff or

developers instead of actual clinicians. IEC 62366-1 requires 'users who are representative of the intended users.'

WW.6 Use-Related Risk Analysis

IEC 62366-1 Clause 5.4 requires identification of hazard-related use scenarios. These are scenarios where the user interface could contribute to a hazardous situation.

Process for use-related risk analysis:

1. Identify user interface elements (displays, controls, labels, audio, etc.)
2. For each element, identify potential use errors
3. For each use error, identify potential hazardous situations
4. Link to hazards in ISO 14971 risk file
5. Determine risk controls in user interface
6. Verify controls through usability testing

Example use-related risk analysis:

Table WW-2: Use-Related Risk Analysis Example

UI Element	Potential Use Error	Hazardous Situation	Hazard (Risk File)	Risk Control
Dose entry field	Enter wrong value	Overdose administered	HA-0015	Confirmation dialog, range check
Patient selector	Select wrong patient	Treatment given to wrong patient	HA-0022	Photo verification, DOB display
Start button	Press unintentionally	Unintended treatment start	HA-0031	Confirmation required, undo option
Alert display	Fail to notice	Delayed response to critical event	HA-0044	Audio alert, visual prominence

The qualitative P1 values shown in Table WW-2 represent initial screening-level estimates appropriate for the early stages of use-related risk analysis described in IEC 62366-1 Clause 5.3. These qualitative estimates serve as a starting point for identifying which use errors warrant deeper quantitative analysis.

For hazards classified as HIGH at CCB, or for use errors identified during formative evaluation that affect safety-critical functions, the qualitative P1 should be decomposed into Pe × P0 using the methodology described in Appendix EEE. This decomposition enables more precise risk control targeting — a qualitative "Occasional" P1 might result from high Pe with low P0 (frequent error, usually caught by the system) or low Pe with high P0 (rare error, always causes hazardous situation), and the appropriate risk control differs between these two cases.

The general principle: use qualitative P1 for initial hazard identification and screening, then decompose to Pe × P0 for safety-critical hazards requiring rigorous analysis. Document both the qualitative screening estimate and the quantitative decomposition in the risk file to show analytical progression.

Appendix XX — AAMI HE75:2025 Human Factors Engineering

AAMI HE75:2025 'Human factors engineering — Design of medical devices' (superseding HE75:2009/(R)2013/(R)2018) provides comprehensive guidance on human factors principles for medical device design. While IEC 62366-1 specifies requirements, HE75 provides the 'how-to' guidance for implementing those requirements. See Appendix KKK for key updates in the 2025 edition.

XX.1 HE75 Overview

HE75 is an ANSI/AAMI standard that provides recommendatory guidance (using 'should' rather than 'shall') with detailed information on:

- Human factors engineering process
- User analysis and task analysis methods
- User interface design principles
- Design of displays, controls, and information presentation
- Testing and evaluation methods
- Documentation requirements

HE75 complements IEC 62366-1 by providing practical guidance on how to achieve usability engineering objectives.

XX.2 User Analysis for SaMD

HE75 Section 6 addresses basic human skills and abilities relevant to user analysis. For SaMD, user analysis should characterize:

User populations:

- Primary users (clinicians who interact directly)
- Secondary users (IT administrators, support staff)
- Patients (if patient-facing features exist)

User characteristics relevant to SaMD:

- Clinical training and expertise

- Technology proficiency
- Physical capabilities (vision, motor control)
- Cognitive load during typical use
- Experience with similar software
- Language and literacy

Use environment characteristics:

- Clinical setting (ICU, clinic, home, etc.)
- Lighting conditions
- Noise levels
- Interruption frequency
- Time pressure
- Device sharing among users

XX.3 Task Analysis

HE75 Section 7 addresses anthropometry and biomechanics. For SaMD, task analysis should leverage these physical interaction principles alongside cognitive task analysis to identify what users need to accomplish and how they accomplish it.

Task analysis methods for SaMD:

- Hierarchical task analysis: Break tasks into subtasks
- Sequential task analysis: Document step-by-step procedures
- Cognitive task analysis: Identify decision points and information needs
- Error analysis: Identify where errors could occur

Example hierarchical task analysis for 'Review patient risk score':

1. Navigate to patient list

1.1 Log into system

1.2 Select patient list view

1.3 Filter or search for patient

2. Select patient

2.1 Verify patient identity

2.2 Click to open patient record

3. Review risk score

3.1 Locate risk score display

3.2 Interpret score value

3.3 Review contributing factors

4. Take action based on score

4.1 Determine if action needed

4.2 Document decision

Task analysis feeds into:

• Requirements (SRS): 'System shall display risk score on patient summary screen'

• Risk analysis: Identify use errors at each step

• Test cases: Verify users can complete tasks

XX.4 User Interface Design Principles

HE75 Sections 15-21 provide detailed guidance on user interface design. Key principles for SaMD:

Display Design (HE75 Section 19)

• Information hierarchy: Most critical information most prominent

• Grouping: Related information grouped together

• Consistency: Similar information displayed consistently

• Legibility: Text readable at intended viewing distance

• Color coding: Meaningful and consistent use of color

Control Design (HE75 Section 18)

- Feedback: Controls provide clear feedback when activated
- Confirmation: Critical actions require confirmation
- Undo: Reversible actions allow undo
- Error prevention: Design prevents errors where possible
- Error recovery: Clear path to recover from errors

Alarm Design (HE75 Section 15)

- Priority levels: Clear distinction between alarm priorities
- Auditory design: Alarms audible in use environment
- Visual design: Visual alarms visible in use environment
- Alarm fatigue mitigation: Avoid excessive or nuisance alarms
- Alarm response: Clear indication of required response

XX.5 HE75 and Agile Development

HE75 guidance can be applied iteratively within sprint-based development:

Table XX-1: HE75 Sprint Integration

HE75 Activity	Sprint Integration	Output
User analysis	Before first sprint (then update)	User profiles in use specification
Task analysis	During requirements refinement	Task flows, use scenarios
Heuristic evaluation	During sprint or at sprint end	Usability findings → Intake
Prototype testing	Within sprint	User feedback → backlog items
Cognitive walkthrough	Before implementation	Design improvements
Usability testing (formative)	At sprint end or between sprints	Findings → Intake → CCB

Key principle: Human factors activities inform design throughout development, not just at the end. Early investment in user analysis and task analysis prevents costly redesign later.

Appendix YY — Usability Engineering Process Walkthrough

This appendix demonstrates how usability engineering integrates with the framework through a practical example.

YY.1 Scenario: New Critical Alert Feature

Product team plans to add a new critical alert feature that notifies clinicians when a patient's risk score exceeds a threshold.

YY.2 Pre-Sprint: Use Specification Development

Before development begins, human factors engineer (or DQE with HF training) develops use specification:

Intended users:

- Primary: Registered nurses, physicians in acute care settings
- Training: Clinical background, 2-4 hours software training
- Technology proficiency: Moderate to high

Use environment:

- Acute care unit with moderate noise levels
- Multiple patients per clinician
- Frequent interruptions
- Shared workstations
- 12-hour shifts with fatigue considerations

User interface elements for this feature:

- Visual alert indicator (color, icon)
- Audio alert (tone)
- Alert message text
- Acknowledge control
- Alert history display

YY.3 Use-Related Risk Analysis

Human factors engineer identifies use-related hazards:

Table YY-1: Human Factors Hazard Analysis

Use Error	Hazardous Situation	Potential Harm	Initial P1	Severity
Fail to notice alert	Clinician unaware of critical condition	Delayed treatment	Occasional	Serious
Dismiss without reading	Critical information not communicated	Inappropriate treatment	Occasional	Serious
Confuse with non-critical alert	Inappropriate prioritization	Delayed response	Probable	Moderate
Unable to locate patient	Delay finding patient after alert	Delayed treatment	Remote	Serious

Note that Table YY-1 uses qualitative P1 estimates (Occasional, Probable, Remote) for initial hazard identification. This is appropriate for the screening phase of use-related risk analysis. However, these qualitative estimates should subsequently be decomposed into Pe and P0 components using the method described in Appendix EEE, particularly for the "Probable" and "Occasional" entries affecting serious harms.

For example, "Dismiss without reading" has P1 = Occasional and Severity = Serious. Decomposing this: Pe might be 20% (based on formative testing showing 1 in 5 users dismiss alerts without reading in high-workload scenarios) and P0 might be 30% (only 30% of dismissed alerts involve critical clinical information in typical alert volumes). This decomposition reveals that risk controls should target both Pe (redesign alert to require active engagement before dismissal) and P0 (reduce alert fatigue so critical alerts are distinguishable from non-critical ones).

The risk controls listed below were identified through this decomposition process, though the table presents only the qualitative summary for readability.

Risk controls identified:

- Multi-modal alerting (visual + audio)
- Distinct alert appearance for critical vs. non-critical
- Patient location clearly displayed in alert
- Require acknowledge action (not auto-dismiss)
- Escalation if not acknowledged within time limit

YY.4 Sprint Development with Usability Activities

Feature enters CCB, classified as HIGH (new clinical alerting feature).

Sprint 1: Design and prototype

- Day 1-2: Requirements refinement with HF input
- Day 3-5: UI design with HF review
- Day 6-7: Prototype development
- Day 8: Heuristic evaluation by HF engineer
- Day 9: Cognitive walkthrough
- Day 10: Design updates based on findings

Formative evaluation findings:

- Alert icon too small for viewing distance → Increased size
- Audio tone similar to other hospital alarms → Changed to distinct tone
- Patient room number not prominent → Moved to larger display area
- Acknowledge button too close to dismiss → Repositioned with separation

Sprint 2: Implementation and formative testing

- Day 1-6: Implementation of designed UI
- Day 7-8: Informal usability testing with 3 clinicians
- Day 9: Updates based on testing

- Day 10: Verification testing

Formative testing findings:

- Users understood alert meaning: 3/3 correct interpretation
- Users located patient: 3/3 within 5 seconds
- One user attempted to dismiss instead of acknowledge → Added 'Are you sure?' confirmation for dismiss

YY.5 Summative Evaluation (Validation Sprint)

After implementation complete, summative evaluation conducted:

Summative evaluation protocol:

- 15 representative users (RNs and physicians from target settings)
- Simulated clinical environment
- Critical tasks defined based on hazard-related use scenarios
- Success criteria defined for each task
- Observer records task completion, errors, time

Critical tasks tested:

- Task 1: Notice and respond to critical alert within 30 seconds
- Task 2: Correctly identify patient and location
- Task 3: Acknowledge alert appropriately
- Task 4: Access alert history

Results:

Table YY-2: Usability Test Results

Task	Success Criterion	Result	Pass/Fail
Notice alert	>90% notice within 30 sec	15/15 (100%)	PASS
Identify patient	>95% correct identification	15/15 (100%)	PASS
Acknowledge	>90% correct action	14/15 (93%)	PASS

Task	Success Criterion	Result	Pass/Fail
Access history	>80% complete within 60 sec	13/15 (87%)	PASS

Residual risk evaluation:

- All critical tasks met acceptance criteria
- One use error observed (acknowledge vs. dismiss confusion) → User self-corrected with confirmation dialog
- Risk controls verified effective
- Residual use-related risk acceptable

YY.6 Documentation in Usability Engineering File

UEF updated with:

- Use specification (updated based on user feedback)
- UI specification (final design)
- Use-related risk analysis (linked to risk file)
- Formative evaluation records (Sprint 1, Sprint 2)
- Summative evaluation protocol and report
- Residual risk evaluation

Cross-references:

- Risk file: HA-0052 through HA-0055 (new hazards)
- DHF: Sprint 1 and Sprint 2 DHF packets
- SRS: Requirements SRS-9.1 through SRS-9.12

> **KEY INSIGHT:** Usability engineering is not a separate activity performed after design—it's integrated throughout. Formative activities during sprints prevent costly late-stage findings. Summative evaluation validates that use-related risks are controlled.

Appendix ZZ — Usability Engineering Regulatory References

This appendix compiles regulatory references for usability engineering requirements.

ZZ.1 IEC 62366-1:2015/AMD1:2020

IEC 62366-1 'Medical devices — Part 1: Application of usability engineering to medical devices'

Key clauses:

- Clause 4: General requirements for usability engineering process
- Clause 5.1: Prepare use specification
- Clause 5.2: Identify user interface characteristics related to safety
- Clause 5.3: Identify known or foreseeable hazards and hazardous situations
- Clause 5.4: Identify and describe hazard-related use scenarios
- Clause 5.5: Select hazard-related use scenarios for summative evaluation
- Clause 5.6: Establish user interface specification
- Clause 5.7: Establish user interface evaluation plan
- Clause 5.8: Design, implement, and perform formative evaluation
- Clause 5.9: Perform summative evaluation
- Clause 5.10: User interface of unknown provenance (UOUP) — alternative evaluation per Annex C

ZZ.2 IEC/TR 62366-2:2016

IEC/TR 62366-2 'Medical devices — Part 2: Guidance on the application of usability engineering to medical devices'

Technical report providing guidance on IEC 62366-1 implementation, including:

- Examples of usability engineering activities
- Guidance on formative and summative evaluation

- Relationship to risk management
- Tailoring the usability engineering effort to project needs

ZZ.3 AAMI HE75:2025 (supersedes HE75:2009/(R)2018)

AAMI HE75 'Human factors engineering — Design of medical devices'

Comprehensive guidance document covering:

- Section 6: Analysis of users and use environments
- Section 7: Task and function analysis
- Section 8: User interface design process
- Section 9: Controls
- Section 10: Visual displays
- Section 11: Auditory signals and displays
- Section 12: Connections and connectors
- Section 13: Software user interfaces
- Section 14: Workstations
- Section 15: Protective equipment
- Section 16: Testing and evaluation
- Section 17: Labeling

ZZ.4 FDA Guidance Documents

FDA guidance on human factors and usability:

'Applying Human Factors and Usability Engineering to Medical Devices' (February 2016)

- Describes FDA expectations for human factors engineering
- Provides guidance on when to submit HFE/usability data
- Describes content of human factors data in submissions

'Human Factors Studies and Related Clinical Study Considerations in Combination Product Design and Development' (February 2016)

- Guidance specific to combination products
- Relevant for SaMD that is part of combination product

ZZ.5 Framework Compliance Summary

Table ZZ-1: Framework Compliance Summary

Requirement Source	Requirement	Framework Element
IEC 62366-1 Clause 4	Usability engineering process	Integrated with design control process
IEC 62366-1 Clause 5.1	Use specification	Part of design input documentation
IEC 62366-1 Clause 5.3	Hazard identification	Risk file with USE-prefix hazards
IEC 62366-1 Clause 5.8	Formative evaluation	Within sprints, findings to Intake
IEC 62366-1 Clause 5.9	Summative evaluation	Validation sprint activity
IEC 62366-1 Clause 5.9 / ISO 14971:2019, 7.3	Residual risk evaluation	Part of risk management report
AAMI HE75 Section 6	User analysis	Use specification development
AAMI HE75 Section 7	Task analysis	Requirements refinement
AAMI HE75 Section 16	Testing and evaluation	Formative and summative testing
FDA Guidance	HFE/usability data	UEF content for submissions

REGULATORY ANCHOR: IEC 62366-1:2015/AMD1:2020 is harmonized with EU MDR and recognized by FDA. AAMI HE75:2025 provides detailed guidance accepted by FDA as demonstrating good human factors practice. Both integrate with ISO 14971 risk management. Organizations should reference the 2025 edition in new usability engineering plans.

Appendix AAA — IEC 62304 Software Lifecycle Integration

IEC 62304:2006/Amd1:2015 (Edition 1.1) 'Medical device software — Software life cycle processes' defines the software development lifecycle requirements for medical device software. This appendix provides comprehensive guidance on integrating IEC 62304 requirements with the framework.

AAA.1 IEC 62304 Overview

IEC 62304 establishes lifecycle requirements for medical device software development and maintenance. The standard applies to software that is a medical device (SaMD) and software that is part of a medical device.

Key concepts:

- Software safety classification determines rigor level
- Risk management integration throughout lifecycle
- Software development process requirements
- Software maintenance process requirements
- Configuration management and problem resolution

AAA.2 Software Safety Classification

IEC 62304 Clause 4.3 requires classification of software systems into safety classes based on the severity of potential harm:

Table AAA-3: IEC 62304 Software Safety Classes

Class	Definition	Potential Harm	Framework Rigor
Class A	No injury or damage to health is possible	None	**LOW rigor path**
Class B	Non-serious injury is possible	Non-serious injury	**MEDIUM rigor path**
Class C	Death or serious injury is possible	Death or serious injury	**HIGH rigor path**

Classification determination:

• Classification is based on the CONTRIBUTION of software to hazardous situations

• If software can contribute to a hazardous situation that could result in death or serious injury → Class C

• If software can contribute only to non-serious injury → Class B

• If software cannot contribute to hazardous situations → Class A

> **REGULATORY ANCHOR**: IEC 62304:2006/Amd1:2015 Clause 4.3 states: 'The MANUFACTURER shall assign a software safety class to each SOFTWARE SYSTEM... based on the possible effects on the patient, user or other person of a HAZARDOUS SITUATION to which the SOFTWARE SYSTEM can contribute.'

AAA.3 Software Safety Class vs. CCB Classification

The framework's CCB classification (HIGH/MEDIUM/LOW) aligns with but is not identical to IEC 62304 software safety class:

Table AAA-4: CCB Classification vs IEC 62304 Class

IEC 62304 Class	Applies To	CCB Classification	Applies To
Class A/B/C	Entire software system	**HIGH/MEDIUM/LOW**	Individual changes
Set once at system level	Based on worst-case harm potential	**Set per change at CCB**	Based on change's safety impact
Determines baseline rigor	For all development	**Determines change rigor**	For that specific change

Relationship:

• A Class C software system may have individual changes classified as MEDIUM or LOW at CCB

• The change doesn't affect safety-critical functionality directly

- But the overall system remains Class C
- Minimum rigor for any change = system safety class requirements

> **KEY INSIGHT:** Software safety class is a system-level classification that sets the floor for rigor. CCB classification is a change-level classification that may require MORE rigor than the floor, but never less.

AAA.4 IEC 62304 Process Requirements by Class

Clause 5: Software Development Process

Table AAA-5 shows the progressive escalation of process requirements from Class A through Class C. The key observation is that Class C requires every activity in the standard, while Class A permits significant streamlining. This maps directly to the framework's rigor levels: a Class C system sets the floor at HIGH rigor, meaning every change must satisfy all Clause 5 requirements regardless of its individual CCB classification.

Table AAA-5: IEC 62304 Clause 5 Requirements by Class

Requirement	Class A	Class B	Class C
5.1 Software development planning	✓ Required	✓ Required	✓ Required
5.2 Software requirements analysis	✓ Required	✓ Required	✓ Required
5.3 Software architectural design	—	✓ Required	✓ Required
5.4 Software detailed design	—	—	✓ Required
5.5 Software unit implementation	✓ Required	✓ Required	✓ Required
5.6 Software integration and testing	—	✓ Required	✓ Required
5.7 Software system testing	✓ Required	✓ Required	✓ Required
5.8 Software release	✓ Required	✓ Required	✓ Required

Framework compliance:

- The framework implements all Class C requirements
- For Class A/B systems, some documentation may be simplified
- CCB classification within a sprint applies additional rigor as needed

Clause 6: Software Maintenance Process

IEC 62304 Clause 6 requires a maintenance process that:

- Establishes criteria for determining when feedback requires modification
- Implements modifications using development process
- Analyzes modifications for impact on safety

Framework compliance:

- PMS signals enter Intake
- CCB classifies impact
- Modifications flow through sprint process
- Risk file updated to reflect impact analysis

Clause 7: Software Risk Management

IEC 62304 Clause 7 requires integration with ISO 14971 risk management:

- Identify software items that could contribute to hazardous situations
- Identify potential causes of contribution
- Document in risk management file
- Implement risk control measures
- Verify risk control measures

Framework compliance:

- Risk file maintained as living document
- DQE updates risk file during sprints
- Risk controls verified through testing
- Traceability from hazards to controls to verification

Clause 8: Software Configuration Management

IEC 62304 Clause 8 requires configuration management that:

- Uniquely identifies configuration items
- Controls changes to configuration items
- Maintains audit trail of changes

Framework compliance:

- Jira provides configuration identification (ticket IDs)
- CCB provides change control
- Jira audit trail captures all changes
- Version control for code and documents

Clause 9: Software Problem Resolution

IEC 62304 Clause 9 requires problem resolution process that:

- Documents problems detected
- Investigates and evaluates problems
- Resolves problems or documents rationale for not resolving
- Verifies resolution

Framework compliance:

- Bugs/defects enter Intake
- CCB classifies severity/priority
- Resolution through sprint process
- Verification confirms fix effectiveness

AAA.5 SOUP and Software of Unknown Provenance

IEC 62304 Clause 5.3.3 and 5.3.4 address SOUP (Software of Unknown Provenance) - software that is already developed and is not part of the device software:

SOUP requirements:

- Identify SOUP items used in the software system
- Document title, manufacturer, version

- Identify and document functional and performance requirements
- Identify known anomalies (published bug lists)
- Evaluate hazardous situations that could result from SOUP failure

For Class B and C software:

- Specify functional and performance requirements for SOUP
- Specify hardware and software necessary for proper operation
- Evaluate SOUP for capability to meet requirements

Framework compliance:

- SOUP inventory maintained as controlled document
- SOUP changes enter Intake and are classified at CCB
- Risk file includes SOUP-related hazards
- Verification includes SOUP integration testing

> **483 PATTERN:** SOUP not identified or evaluated. Organization uses open-source libraries but has no inventory, no anomaly review, and no risk evaluation. IEC 62304 requires identification and evaluation of all SOUP items.

AAA.6 Software Unit Verification

IEC 62304 Clause 5.5.5 requires verification of software units:

For Class C software:

- Verify each unit meets detailed design
- Establish acceptance criteria before verification
- Document verification results

Framework compliance:

- Unit tests verify unit functionality
- Test coverage metrics demonstrate verification
- DQE reviews verification adequacy

AAA.7 Traceability Requirements

IEC 62304 requires traceability at multiple levels:

Table AAA-6: IEC 62304 Traceability Requirements

From	To	IEC 62304 Reference	Framework Element
System requirements	Software requirements	5.2.6	SRS linkage
Software requirements	Software architecture	5.3.6	SDS linkage
Software architecture	Detailed design	5.4.4	Design documentation
Detailed design	Software units	5.5.5	Code comments/links
Software requirements	System tests	5.7.4	Test case links
Software requirements	Risk control measures	7.2	Risk file links

The framework's Jira-based traceability satisfies these requirements through linked fields and traceability reports.

Appendix BBB — Software Development Plan Template

IEC 62304 Clause 5.1 requires a software development plan. This appendix provides a template outline.

BBB.1 Software Development Plan Contents

1. Introduction
   - 1.1 Purpose and scope
   - 1.2 Definitions and abbreviations
   - 1.3 References
2. Software Description
   - 2.1 Software system overview
   - 2.2 Software safety classification
   - 2.3 Intended use and users
3. Development Process
   - 3.1 Software development lifecycle model (Agile/Scrum)
   - 3.2 Sprint structure and cadence
   - 3.3 CCB and change control process
   - 3.4 Definition of Done
4. Deliverables
   - 4.1 Software requirements specification (SRS)
   - 4.2 Software architecture (SDS)
   - 4.3 Software detailed design
   - 4.4 Source code
   - 4.5 Verification and test documentation
   - 4.6 Software release package

5. Traceability
   - 5.1 Traceability approach
   - 5.2 Traceability tools (Jira configuration)
6. Configuration Management
   - 6.1 Version control
   - 6.2 Build management
   - 6.3 Release management
7. Problem Resolution
   - 7.1 Bug/defect handling
   - 7.2 Escalation criteria
8. Risk Management Integration
   - 8.1 Reference to risk management plan
   - 8.2 Risk activities in development
9. Verification
   - 9.1 Unit verification approach
   - 9.2 Integration verification approach
   - 9.3 System verification approach
10. Documentation Control
    - 10.1 Document identification
    - 10.2 Review and approval
    - 10.3 Change control

11. SOUP Management
   - 11.1 SOUP identification process
   - 11.2 SOUP evaluation criteria
   - 11.3 SOUP anomaly review
12. Tools
   - 12.1 Development tools
   - 12.2 Test tools
   - 12.3 Tool validation

REGULATORY ANCHOR: IEC 62304:2006/Amd1:2015 Clause 5.1.1 requires: 'The MANUFACTURER shall establish a SOFTWARE DEVELOPMENT PLAN... for conducting the activities of the software development process appropriate to the scope, magnitude, and SOFTWARE SAFETY CLASSIFICATIONS of the SOFTWARE SYSTEM to be developed.'

Appendix CCC — Operational Role Clarity: De-Coupling QA/RA

A common failure mode in regulated development is conflating 'QA/RA' into a single entity. While conceptually related, Quality and Regulatory have fundamentally different operational cadences and responsibilities. This appendix provides detailed guidance on operational role separation.

CCC.1 The QA/RA Conflation Problem

Many organizations treat 'QA/RA' as a single function. This creates several problems:

Bottleneck creation: When RA is required for sprint-level decisions, releases stall waiting for RA availability. RA specialists are typically shared resources who cannot attend every standup or sprint planning session.

Role confusion: When 'QA/RA' approves something, it's unclear whether Quality approved the verification or Regulatory approved the submission impact. Both may be required, but they're different approvals.

Cadence mismatch: Quality activities can happen at sprint cadence. Regulatory activities typically happen at milestone cadence. Forcing both into the same cadence either slows Quality or overwhelms Regulatory.

> **KEY INSIGHT:** The framework's DQE role exists specifically to solve this problem. DQE provides sprint-cadence quality involvement without requiring RA to attend every sprint ceremony. RA remains focused on submission-level strategy.

CCC.2 Three Distinct Quality/Regulatory Functions

The framework distinguishes three functions that are often conflated:

Quality Assurance (QA) — The Auditor
Scope: Process compliance and audit

Cadence: Audit cadence (quarterly, annual)

Focus: Is the organization following its procedures?

Activities: Internal audits, CAPA management, supplier audits, management review

Not involved in: Sprint-level decisions, daily verification

Design Quality Engineer (DQE) — The Embedded Partner
Scope: Sprint-level execution and verification

Cadence: Sprint cadence (daily/weekly)

Focus: Is this specific change properly controlled?

Activities: CCB classification, verification approval, risk file updates during sprints, traceability validation

Key distinction: DQE is embedded in the sprint team, not external to it

Regulatory Affairs (RA) — The Strategic Gatekeeper
Scope: Submission strategy and regulatory compliance

Cadence: Milestone cadence (release, submission)

Focus: Does this require regulatory action?

Activities: Submission impact assessment, predicate analysis, agency communications, labeling review

Key distinction: RA is consulted on HIGH items but not required for every sprint decision

Table CCC-1: QA vs DQE vs RA Function Comparison

Function	QA	DQE	RA
Cadence	Audit (quarterly)	Sprint (daily/weekly)	Milestone (release)
Primary question	Are we following process?	Is this change controlled?	Does this need submission?
Sprint involvement	None	Full participation	Consulted on HIGH items
Release involvement	Audit readiness	Verification approval	Submission sign-off
CCB role	None	Classification owner	Consulted on impact
Bottleneck risk	Low	Low (embedded)	High if over-involved

The most important pattern in this table is cadence separation. QA operates at audit cadence—quarterly or event-driven. DQE operates at sprint cadence—daily and weekly. RA operates at milestone cadence—release gates and submissions. When organizations collapse these into a single "QA/RA" function, the result is predictable: either the combined role operates at sprint cadence (neglecting audit readiness and regulatory strategy) or at milestone cadence (creating the bottleneck that the framework is designed to eliminate).

The DQE role is the framework's key innovation. By embedding quality engineering at sprint cadence, classification and verification happen in real time rather than as retroactive audits. This is what makes the framework native rather than retrofit—quality decisions are made when context is fresh and course corrections are cheap.

CCC.3 The Traffic Controllers: PM and Program Manager

Agile requires rhythm; compliance requires rigorous checkpoints. Project and Program Managers manage this tension.

Project Manager (PM) — The Compliance Police
The PM enforces the daily discipline of the regulated sprint. They are the 'compliance police' of the sprint.

Workflow enforcement responsibilities:

• Ensures no ticket moves to 'In Progress' without Safety_Impact classification

• Ensures no ticket moves to 'Done' without required evidence links

• Facilitates CCB meetings and ensures classification happens before sprint entry

• Maintains tracker accuracy and escalates blockers

• Monitors documentation lag and raises alarms when evidence falls behind

Daily standup compliance questions:

• 'Is your documentation keeping pace with your code?'

• 'Are there any risk implications from yesterday's work?'

• 'Is the tracker current?'

> **ROLE ACCOUNTABILITY:** PM is the process enforcer, not the process owner. PM ensures the rules are followed but doesn't make classification or approval decisions. When a developer tries to skip CCB, PM blocks the attempt. When evidence is missing at sprint end, PM escalates.

Program Manager — The Release Orchestrator
The Program Manager operates at release level, coordinating cross-functional activities.

Release-level responsibilities:

• Owns the Validation Sprint coordination

• Secures resources for validation (clinical staff, test environments)

- Schedules Release Gate reviews
- Compiles Technical File for audit readiness
- Makes the release decision based on DQE/RA approvals

Cross-functional coordination:

- Manages dependencies between product teams
- Coordinates with RA on submission timing
- Ensures PMS readiness for release

> **ROLE ACCOUNTABILITY:** Program Manager is the only role that can approve release, but they do so based on evidence that DQE and RA have approved. Program Manager doesn't override DQE/RA decisions—they confirm the decisions have been made.

CCC.4 The Signal Processors: Complaints and PMS

Field signals require two distinct processing functions that are often conflated:

Complaints Specialist — The Triage Function

The Complaints Specialist receives raw signals and determines their nature.

Responsibilities:

- Receives raw input (calls, emails, field reports)
- Determines: 'Is this a complaint per regulatory definition?'
- Documents complaint per 21 CFR 820.198 requirements
- Feeds the Intake stream with documented complaints

What Complaints Specialist does NOT do:

- Does not determine if design change is needed
- Does not classify severity or priority

• Does not trigger CAPA (that's PMS Manager)

• Does not analyze trends (that's PMS Manager)

PMS Manager — The Trend Analyst
The PMS Manager analyzes aggregate data and triggers action.

Responsibilities:

• Analyzes complaint trends over time

• Monitors for signals that cross thresholds (e.g., P1 moving from 'Remote' to 'Occasional')

• Triggers CAPAs when patterns emerge

• Creates Intake signals that require design investigation

• Updates risk file probability estimates based on field data

• Monitors effectiveness of corrective actions

Key distinction: Complaints Specialist processes individual signals. PMS Manager sees the forest, not just the trees.

> **ROLE ACCOUNTABILITY:** Complaints Specialist feeds Intake. PMS Manager analyzes what's in Intake and triggers action. When a trend emerges, PMS Manager creates the Intake item that forces Engineering to respond. The separation prevents both under-reaction (missing trends) and over-reaction (treating every complaint as a design issue).

CCC.5 Enhanced RACI: Operational Clarity

This RACI matrix emphasizes the cadence and accountability distinctions:

Table CCC-2: Framework RACI with Cadence

Activity	Eng	DQE	RA	PO	PM	PgM	Complaints	PMS
Intake (Feature)	C	I	C*	R	I	I	—	—
Intake (Field Signal)	I	C	C	I	I	I	R	A
CCB Classification	C	A	C*	C	R	I	—	C
Sprint Planning	R	C	I	A	R	I	—	—
Risk File Update	C	A	I	I	I	I	—	C
Implementation	A	I	I	C	I	I	—	—
Verification	R	A	I	I	I	I	—	—
Validation Sprint	R	C	C	R	R	A	—	—
Release Gate	I	A	A	C	R	A	—	I
Complaint Triage	—	I	I	—	—	—	A	C
Trend Analysis	—	C	I	—	—	—	C	A
CAPA Trigger	—	C	I	—	—	C	I	A
SOUP Register Update	R	A	I	I	I	I	—	—
UEF Maintenance	C	A	I	C	I	I	—	—
EU Tech Doc Compilation	C	R	A	C	R	I	—	—
GSPR Mapping	C	R	A	I	I	I	—	—
PMS Plan/PSUR (EU)	—	C	C	—	—	—	C	A
NB Audit Preparation	C	R	A	I	R	I	—	C
Safety Class Determination	C	A	C	I	I	I	—	—
PCCP Monitoring (AI/ML)	C	R	A	C	I	I	—	C

A **Accountable** — *Owns the outcome (single person)*
R **Responsible** — *Does the work*
C **Consulted** — *Provides input before decision*
I **Informed** — *Kept aware after decision*

Key patterns in this RACI:

• RA is Consulted at Intake/CCB but Accountable only at Release Gate — prevents RA bottleneck

• DQE is Accountable for Classification, Risk File, and Verification — sprint-level quality ownership

• PM is Responsible for CCB facilitation and Release packaging — process enforcement

• PgM is Accountable for Validation Sprint and Release Gate — release-level ownership

• Complaints and PMS have distinct accountabilities — signal processing separation

For organizations operating in EU markets, additional RACI rows apply: RA is Accountable for Technical Documentation, GSPR mapping, and Notified Body audit coordination. PMS Manager is Accountable for PMS Plan, PMS Report/PSUR, and trend analysis feeding into PSUR content. DQE is Accountable for SOUP register oversight, IEC 62304 safety class determination, and UEF integration with the risk file. These extend the core RACI without changing its fundamental structure — the same cadence separation principles apply.

> **483 PATTERN:** QA/RA conflation creates bottleneck. Organization requires 'QA/RA approval' for every change. RA cannot keep up with sprint cadence. Result: either RA becomes overwhelmed, or teams bypass the approval. Solution: DQE handles sprint-level approval; RA handles release-level and HIGH-item consultation.

Appendix DDD — PM Workflow Enforcement Rules

This appendix provides specific workflow enforcement rules for the Project Manager role.

DDD.1 Jira Transition Rules

PM ensures these workflow rules are configured and enforced:

Backlog → CCB_Queue
Trigger: Item prioritized for upcoming sprint

Validation: None required to enter queue (anyone can nominate), but sprint entry is blocked until CCB classification is complete

PM action: Ensure CCB agenda includes all queued items

CCB_Queue → Classified
Trigger: CCB classification complete

Validation required:

- Safety_Impact field is not empty (HIGH/MEDIUM/LOW)

- CCB_Reference field contains link to CCB record

PM action: Reject transition if validation fails

Classified → In_Progress
Trigger: Item pulled into sprint

Validation required:

- Safety_Impact is set

- Sprint is assigned

PM action: Block 'In Progress' for unclassified items

In_Progress → In_Review
Trigger: Implementation complete

Validation required:

- For MEDIUM/HIGH: Risk_Link is not empty

- For MEDIUM/HIGH: SRS_Link is not empty

PM action: Remind developer of evidence requirements

In_Review Decision (Blocked?)
Trigger: Verification complete

Validation required:

• For HIGH: Verification_Status = APPROVED AND RA_Review = COMPLETE

• For MEDIUM: Verification_Status = APPROVED

• For LOW: Verification_Status = PASSED or N/A

• All required links populated

PM action: Block 'Done' until all evidence complete

DDD.2 Daily Standup Compliance Checks

PM performs these checks at each standup:

Documentation currency check:

• 'For items In_Progress more than 2 days, is documentation current?'

• Flag items where code is ahead of documentation

Classification check:

• 'Are there any items In_Progress without Safety_Impact?'

• These should not exist if workflow is enforced

Evidence check:

• 'For items moving to In_Review today, are evidence links ready?'

• Prevent end-of-sprint evidence scramble

DDD.3 Sprint Boundary Enforcement

PM enforces these rules at sprint boundaries:

Sprint Start

• All sprint items have Safety_Impact classification

• HIGH items have RA consultation scheduled if needed

• DQE availability confirmed for sprint

Sprint End

- All 'Done' items have complete evidence per classification
- Carryover items documented with reason
- Sprint DHF packet generated
- Metrics captured (evidence completion rate, documentation lag)

Release Gate

- All items in release have required approvals
- Tracker shows no BLOCKED items
- DQE confirms verification complete
- RA confirms submission assessment complete (for HIGH items)
- Program Manager reviews and approves

> **ROLE ACCOUNTABILITY:** PM is empowered to block transitions that don't meet criteria. This is not bureaucracy—it's prevention. Blocked transitions reduce the likelihood of common 483 patterns: documentation lag, missing approvals, incomplete evidence. PM should be supported by leadership in enforcing these rules.

Appendix EEE — Quantitative P1/P2 Probability Estimation

ISO 14971 defines risk as the combination of probability of occurrence of harm and severity of that harm. Probability of harm (Pharm) is given by P1 × P2, where P1 is probability of hazardous situation occurring and P2 is probability of hazardous situation leading to harm. This appendix provides a method for quantitative estimation of these probabilities, particularly for use-related risks.

EEE.1 The Probability Estimation Challenge

A key challenge in risk analysis is that probability estimates often rely on subjective judgment. Stakeholders may vary significantly in their estimates. The ISO 14971 framework provides the structure (P1 × P2) but doesn't specify how to derive these estimates quantitatively.

Common problems with probability estimation:

- Estimates are inconsistent across assessors
- 'Occasional' or 'Remote' categories lack precision
- Use error probability is conflated with harm probability
- Updates based on field data are difficult to incorporate

EEE.2 Decomposing P1: The Pe × P0 Method

P1 (probability of hazardous situation occurring) can be challenging to estimate directly. A more tractable approach decomposes P1 into two constituent probabilities:

Pe = Probability of error occurring

The likelihood that a user commits a specific use error. This can be estimated from usability studies, literature, field data, or expert judgment.

P0 = Probability of error causing the hazardous situation

The likelihood that, given the error occurred, it results in the hazardous situation. Not all errors lead to hazardous situations—P0 captures this conditional relationship.

Therefore: $P1 = Pe \times P0$

And the complete probability chain becomes:

$Pharm = Pe \times P0 \times P2 = P1 \times P2$

> **KEY INSIGHT:** Breaking P1 into Pe and P0 makes estimation more tractable. Pe can often be measured empirically through usability testing. P0 requires analysis of whether the error actually creates the hazardous situation. This decomposition improves both accuracy and traceability.

EEE.3 The Probability Chain Model

The probability chain follows the Swiss Cheese Model principle—harm occurs only when failures align across multiple barriers:

Error → Hazard → Hazardous Situation → Harm

Each transition has an associated probability:

Table EEE-1: Probability Symbol Definitions

Symbol	Definition	Example
Pe	Probability of error occurring	User fails to check expiry date (50%)
P0	Probability of error causing hazardous situation	Expired product is actually present (5%)
P1	Probability of hazardous situation (Pe × P0)	User administers expired product (2.5%)
P2	Probability of hazardous situation → harm	Expired product causes adverse effect (0.01%)
Pharm	Overall probability of harm (P1 × P2)	Harm from expired product (0.00025%)

EEE.4 Worked Example 1: Low P2 Scenario

Scenario: User fails to wash hands before injection

Table EEE-2: Pharm Calculation Example 1

Component	Definition	Value	Rationale
Pe	Probability of error	50%	Usability studies show 50% skip handwashing
P0	Error → hazardous situation	5%	Only 5% of unwashed hands contaminate site
P1	Hazardous situation occurs	2.5%	Pe × P0 = 0.5 × 0.05
P2	Hazardous situation → harm	0.01%	Very low probability of systemic infection
Pharm	Overall harm probability	0.00025%	P1 × P2 = 0.025 × 0.0001

Interpretation: Although the use error is common (Pe = 50%), the overall harm probability is very low because P0 and P2 are both low. The barriers between error and harm are effective.

EEE.5 Worked Example 2: High P2 Scenario

Scenario: User lifts injection device before dose completion

Table EEE-3: Pharm Calculation Example 2

Component	Definition	Value	Rationale
Pe	Probability of error	10%	Usability studies show 10% lift early
P0	Error → hazardous situation	100%	Early lift always causes incomplete dose
P1	Hazardous situation occurs	10%	Pe × P0 = 0.1 × 1.0
P2	Hazardous situation → harm	80%	High probability of clinical impact from underdose
Pharm	Overall harm probability	8%	P1 × P2 = 0.1 × 0.8

Interpretation: Although the use error is less common (Pe = 10%), the overall harm probability is high because when the error occurs, it reliably causes harm (P0 = 100%, P2 = 80%). This scenario requires risk control.

> **KEY INSIGHT:** High Pe does not mean high risk. Low Pe does not mean low risk. The complete chain (Pe × P0 × P2) determines risk. Focus risk controls where the chain is strongest, not just where errors are most common.

EEE.6 Sources for Probability Estimates

Pe (Error Probability) Sources

- Formative usability studies (observed error rates)
- Summative usability studies (validated error rates)
- Literature on similar devices/interfaces
- Field complaint data from similar products
- Human reliability analysis databases (e.g., THERP)
- Expert judgment (when data unavailable)

P0 (Error → Hazardous Situation) Sources

- Engineering analysis of failure modes
- Fault tree analysis
- Historical data on error consequences
- Expert judgment based on device design

P2 (Hazardous Situation → Harm) Sources

- Clinical literature on condition severity
- Clinical expert judgment
- Epidemiological data
- Post-market surveillance data
- Clinical trial adverse event data

Evidence Hierarchy and Source Selection

Not all sources carry equal evidentiary weight. When multiple sources are available, the risk file should document which source was used and why it was selected over alternatives. The general hierarchy, from strongest to weakest evidence:

For Pe: Summative usability study data (device-specific, validated sample) is strongest, followed by formative usability study data (device-specific, exploratory sample), then field data from substantially similar predicates, published literature for analogous device types, human reliability databases (generic task categories), and finally expert judgment.

For P0: Deterministic engineering analysis (fault tree, FMEA showing causal path) is strongest, followed by historical field data showing error-to-consequence relationship, then expert judgment based on device architecture review.

For P2: Epidemiological data for the specific clinical scenario is strongest, followed by clinical trial adverse event data, published clinical literature, and clinical expert judgment.

When sources conflict — for example, when formative usability testing shows a 5% error rate but published literature for analogous devices shows 15% — document both values, explain the discrepancy (sample size, population differences, task fidelity, environmental differences), and state which value was selected for the risk estimate and why. Conservative selection (using the higher Pe) is defensible and demonstrates precautionary thinking, but overly conservative estimates across all hazards can make the risk file implausible and may cause auditors to question the analytical rigor.

When no empirical data exists, expert judgment is acceptable but must be documented with the expert's qualifications, the basis for the estimate, and any assumptions. A statement such as "Pe estimated at 10% based on DQE judgment informed by 15 years of experience with similar numeric entry interfaces in clinical settings" is substantially more defensible than "Pe = Remote."

EEE.7 Converting Quantitative to Qualitative

Risk matrices typically use qualitative categories. Map quantitative Pharm to qualitative categories:

Table EEE-4: Pharm to Qualitative Category Mapping

Pharm Range	Qualitative Category	Typical Score
> 10%	Frequent	6
1% - 10%	Probable	5
0.1% - 1%	Occasional	4
0.01% - 0.1%	Remote	3
0.001% - 0.01%	Improbable	2
< 0.001%	Incredible	1

These thresholds should be defined by the organization and documented in the risk management plan. The key benefit of quantitative estimation is that the mapping is explicit and traceable.

Establishing and Calibrating Thresholds

The threshold values in Table EEE-4 are illustrative — each organization must define its own mapping based on the device type, intended use environment, and risk acceptability policy documented in its risk management plan (ISO 14971:2019 Clause 4.4). A Class III implantable device may warrant tighter thresholds (lower Pharm values triggering higher-risk categories) than a Class I wellness application.

The calibration process should involve DQE, RA, and clinical stakeholders reviewing the proposed thresholds against the organization's existing risk acceptance matrix. If the organization already has an established qualitative matrix, the quantitative thresholds should be set so that the mapping produces results consistent with historical risk assessments — cases that were previously classified as "Occasional" based on qualitative judgment should fall within the "Occasional" range of the quantitative mapping. Significant discrepancies indicate either that the thresholds need adjustment or that historical qualitative assessments were miscalibrated — both findings are valuable.

Document the threshold selection rationale in the risk management plan. Auditors may ask why the boundary between “Probable” and “Occasional” is set at 1% rather than 5%. The answer should reference device-specific clinical context, not arbitrary convention.

EEE.8 Updating Estimates with Field Data

A major advantage of quantitative estimation is the ability to update estimates when field data becomes available:

Before market: Pe estimated from usability studies (e.g., 10%)

After 1 year: Field complaints suggest actual Pe is 15%

Action: Update Pe in risk file, recalculate Pharm, reassess acceptability

This enables evidence-based updates to the risk file as post-market data accumulates. PMS Manager monitors for signals that would trigger such updates.

The update process is not a one-time correction — it is an ongoing capability that supports the continuous improvement loop central to this framework.

When to Update

Not every field complaint warrants a Pe revision. The PMS Manager should apply signal detection methodology to distinguish between isolated events and genuine trends. Triggers for Pe re-estimation include: complaint rates for a specific use error exceeding the pre-market Pe estimate by a clinically meaningful margin, usability-related adverse events reported to FDA (MDR) or through the organization’s vigilance system, post-market usability studies revealing use patterns not anticipated during pre-market testing, or changes in the user population or use environment that affect error probability (e.g., expanded indications introducing new user groups).

How to Update

When field data warrants a Pe update, the process involves several steps. First, calculate the observed field error rate from complaint data, normalizing against the appropriate denominator (total uses, total users, total device-hours — whichever matches the Pe definition in the risk file). Second, compare the observed rate to the pre-market estimate and document the magnitude and direction of the discrepancy. Third, determine whether the discrepancy changes the qualitative risk category using the organization's Pharm-to-qualitative mapping. Fourth, if the risk category changes, evaluate whether additional risk controls are needed — this evaluation loops back through the CCB process as a new change item. Fifth, update the risk file with the revised Pe, the field data source and sample size, the date of the revision, and the rationale for the new estimate.

Documentation Rigor

The updated risk file entry should show the progression of Pe estimates over time: the initial estimate with its source, each revision with its triggering data, and the current estimate. This history demonstrates exactly the kind of evidence-based risk management that ISO 14971:2019 Clause 10 (production and post-production activities) requires. An auditor reviewing the risk file should be able to trace the evolution of a probability estimate from pre-market hypothesis through post-market validation — this is the gold standard for demonstrating that risk management is truly ongoing.

Common Pitfall: Small Sample Bias

Field complaint data is inherently biased toward under-reporting. Not all use errors produce complaints — many are caught by the user, corrected, and never reported. This means that observed complaint rates typically underestimate actual error rates. When updating Pe from field data, document the assumed reporting fraction and consider whether the observed rate should be adjusted upward. Conversely, early post-market data from small installed bases may show artificially high error rates due to the learning curve effect — error rates often decrease as users become more experienced with the device. Document these considerations alongside the estimate to demonstrate analytical sophistication.

EEE.9 Integration with Framework

The quantitative estimation method integrates with the framework at several points:

At CCB: DQE considers Pe estimates when classifying changes. Changes that increase Pe should be classified higher.

During Sprint: When updating risk file, DQE documents Pe, P0, P1, and P2 estimates with rationale and sources.

At Release Gate: Risk file includes quantitative estimates for key hazards, demonstrating analytical rigor.

In PMS: PMS Manager compares field data to Pe estimates. Significant deviations trigger risk file updates.

> **REGULATORY ANCHOR:** ISO 14971:2019 Clause 5.5 requires estimation of risk for each hazardous situation. The standard notes that probability 'can be estimated using available data or generated data' and that 'the estimation of probability can use sources including published standards, scientific technical data, field data from similar medical devices, and usability tests.'

> **483 PATTERN:** Probability estimates lack rationale. Risk file shows 'Remote' probability but no explanation of how this was determined. Inspector asks 'What data supports this estimate?' No answer available. Quantitative estimation with documented sources prevents this finding.

Appendix FFF — Use-Related Risk Analysis: HF and Risk Integration

IEC 62366-1 requires identification of use-related hazards. ISO 14971 requires risk analysis. This appendix provides guidance on integrating use-related risk analysis with the quantitative probability method.

FFF.1 The Use Error Problem

A common mistake in risk analysis is treating 'user error' as a hazard. This is incorrect per ISO 14971 definitions:

WRONG: Hazard = 'User enters wrong dose'

RIGHT: Hazard = 'Dose entry field accepts values outside safe range'

The user behavior (entering wrong dose) appears in the hazardous situation, not the hazard. The hazard is the device characteristic that enables the harm.

Correct causal chain:

- Hazard (device characteristic): Dose entry field accepts any numeric value
- Sequence of events: User intends 10mg, types 100mg, doesn't notice error
- Hazardous situation: 100mg dose prepared for administration
- Harm: Overdose with clinical consequences

FFF.2 Mapping Use Errors to Probability Components

The Pe × P0 × P2 method maps directly to use error analysis:

Table FFF-5: Use Error Analysis to Probability Mapping

Use Error Analysis	Probability Component	Description
Use error identified	Pe	How likely is this error?
Error → hazardous situation	P0	If error occurs, does it create hazard?
Hazardous situation probability	P1 = Pe × P0	Combined probability of exposure
Hazardous situation → harm	P2	Clinical probability of harm

Pe Estimation: The Source Hierarchy

The table states that Pe captures "how likely is this error?" but does not address where Pe values come from — a question that auditors routinely ask and that ISO 14971:2019 Clause 5.5 explicitly addresses by noting that probability "can be estimated using available data or generated data." Pe estimation sources form a hierarchy of evidentiary strength, and the appropriate source depends on the product lifecycle stage.

Pre-design (concept phase): Pe estimates rely on published literature for analogous devices, human reliability analysis databases (e.g., THERP, CREAM), and expert judgment. At this stage, estimates are inherently uncertain, and the risk file should document this uncertainty explicitly — stating "Pe estimated at 15% based on published error rates for free-text numeric entry in clinical settings (Smith et al., 2019)" rather than simply "Pe = 15%." The rationale is as important as the number.

Formative evaluation: Observed error rates from formative usability studies provide empirical Pe estimates, but with important caveats. Formative studies typically use small samples (5–8 participants per user group) in controlled conditions. Error rates observed in formative testing may underestimate or overestimate actual field error rates depending on task fidelity, participant representativeness, and environmental simulation quality. Document these limitations alongside the estimates.

Summative validation: Validated error rates from summative usability studies with statistically justified sample sizes provide the strongest pre-market Pe estimates. IEC 62366-1 Clause 5.9 requires summative evaluation for hazard-related use scenarios. These studies produce Pe estimates with known confidence intervals — document both the point estimate and the confidence interval in the risk file.

Post-market surveillance: Field complaint data, adverse event reports, and post-market usability studies provide real-world Pe values. These data may differ significantly from pre-market estimates because actual use environments, user populations, and use patterns are more diverse than what testing captures. When field data becomes available, update Pe estimates and recalculate Pharm — this is a PMS Manager responsibility that triggers DQE risk file review (see EEE.8).

The practical implication is that Pe is not a static number assigned at design input and forgotten. It is a living estimate that should evolve through the product lifecycle, with each revision documented along with its source and rationale. This lifecycle progression — from literature-based estimate to formative observation to summative validation to field confirmation — should be visible in the risk file's revision history.

P0 Behavior for Use-Related Risks

P0 — the probability that an error causes the hazardous situation — behaves differently for use errors than for hardware or software failures. For a software defect, P0 is often deterministic: a specific input always produces a specific output, so P0 is either 1.0 or 0.0 depending on whether the fault condition leads to the hazardous situation. For use errors, P0 depends on whether the system includes error-tolerant design features — and in the absence of such features, P0 defaults to 1.0.

This default is important because it represents a common auditor discussion point. If a dose entry field accepts any numeric value and the user enters the wrong value, P0 = 1.0: the error always produces the hazardous situation (incorrect dose prepared). The system has no mechanism to intercept the error between occurrence and consequence. When documenting P0 = 1.0, the rationale should state explicitly that no error-trapping mechanism exists for this failure mode — because the natural follow-up question from an auditor is "then why haven't you implemented one?"

Conversely, P0 < 1.0 for use errors indicates that the system has some inherent error tolerance. If the dose entry field accepts any value but the system flags entries outside the 5th–95th percentile range for confirmation, then some erroneous entries will be caught by the confirmation step. P0 in this case reflects the probability that the user dismisses or ignores the confirmation — which itself is a use error that may warrant its own analysis. This recursive quality of use-error analysis (where a risk control introduces new use error modes) is characteristic of human factors engineering and should be documented in the risk file.

Foreseeable Misuse vs. Normal Use Error

IEC 62366-1 distinguishes between use errors occurring during normal use and foreseeable misuse. Both must be analyzed, but they require different Pe estimation approaches.

Normal use errors are errors that occur when the user is attempting to use the device as intended. These are the errors typically captured in usability testing: selecting the wrong patient from a list, misreading a value, failing to notice an alert. Pe for normal use errors can be estimated directly from usability study observations because the testing scenario matches the intended use scenario.

Foreseeable misuse encompasses reasonably foreseeable use that is not intended by the manufacturer but that can be anticipated based on knowledge of human behavior. Examples include using a device designed for single-patient monitoring to display data for multiple patients simultaneously, or employing workarounds that bypass intended workflows. Pe for foreseeable misuse cannot be reliably estimated from standard usability testing because the testing protocol is designed around intended use. Instead, Pe estimates for foreseeable misuse typically derive from field complaint analysis of predicate or analogous devices, published literature on workaround behaviors in the target use environment, contextual inquiry observations (where researchers observe actual clinical practice rather than testing scripted scenarios), and expert judgment from clinicians familiar with the target workflow.

The distinction matters operationally because foreseeable misuse often produces P0 values closer to 1.0. When a user is employing an unintended workflow, the system's error-tolerant design features — which were designed around intended use — may not function as expected. A confirmation dialog that protects against accidental wrong-patient selection during intended use may be routinely dismissed by users who have developed a workaround involving rapid cycling through patient records.

Document the normal-use vs. foreseeable-misuse distinction in the risk file. For each use error, indicate whether it represents normal use or foreseeable misuse, and cite the appropriate evidence source for the Pe estimate. This documentation supports the IEC 62366-1 Clause 5.3 requirement to identify "known or foreseeable hazards and hazardous situations" and demonstrates that the analysis considered the full spectrum of reasonably foreseeable use.

FFF.3 Risk Control Targeting

Risk controls can target different points in the probability chain:

Reducing Pe (Prevent the Error)

- Improve user interface design
- Add forcing functions
- Improve labeling and instructions
- Provide training

Example: Add dropdown with valid dose options instead of free-text entry

Reducing P0 (Error Doesn't Cause Hazardous Situation)

- Add validation/verification steps
- Implement automatic checks

• Add confirmation dialogs

Example: System validates dose against safe range before accepting

Reducing P2 (Hazardous Situation Doesn't Cause Harm)

• Add detection mechanisms

• Implement recovery options

• Provide alerts/warnings

Example: Alert displayed before administration if dose exceeds typical range

> **KEY INSIGHT:** The most effective risk controls reduce Pe (prevent errors through design). The next most effective reduce P0 (errors are caught before causing hazardous situation). Reducing P2 is last resort—it means the hazardous situation already occurred.

FFF.4 Documenting Use-Related Risk Analysis

For each use-related hazard, document:

1. User task and interface element involved
2. Potential use error (what could go wrong)
3. Pe estimate with source (usability data, literature, judgment)
4. Hazardous situation if error occurs
5. P0 estimate with rationale
6. Resulting $P1 = Pe \times P0$
7. Potential harm
8. P2 estimate with clinical rationale
9. Resulting $Pharm = P1 \times P2$
10. Severity estimate
11. Risk evaluation (acceptable/unacceptable)
12. Risk controls if needed
13. Residual risk after controls

FFF.5 Example: Complete Use-Related Risk Entry

Table FFF-6: Complete Use-Related Risk Entry Example

Field	Content
Hazard ID	HA-0089
User task	Enter patient weight for dose calculation
UI element	Weight entry field (free text, kg)
Use error	User enters weight in lbs instead of kg
Pe	15% (based on formative usability study with US users)
Hazardous situation	Dose calculated based on incorrect weight
P0	100% (system has no unit validation)
P1	15% (0.15 × 1.0)
Harm	Overdose (if lbs > kg) or underdose (if kg interpreted as lbs)
P2	70% (calculated dose would be significantly wrong)
Pharm	10.5% (0.15 × 0.70)
Severity	Serious (potential for significant clinical harm)
Initial risk	Unacceptable (Pharm 10.5%, Severity Serious)
Risk control	RC-0089: Add unit selector (kg/lbs) with default based on locale
Post-control Pe	2% (validated in follow-up usability study)
Residual Pharm	1.4% (0.02 × 1.0 × 0.70)
Residual risk	Acceptable (reduced probability with unchanged severity)

This level of documentation demonstrates analytical rigor and provides clear traceability from use error through hazardous situation to risk control to residual risk.

Framework Overview

Before diving into the details, this diagram provides a complete visual overview of the Regulated Sprint Process. It shows how all components—Intake, CCB, Sprint execution, verification, release gate, and post-market surveillance—integrate into a unified FDA and EU MDR compliant workflow.

Study this diagram carefully. Every element described in the subsequent chapters appears here. The color coding indicates:

- GREEN: Inputs and Governance (Intake, CCB)
- BLUE: Execution and Verification (Sprint activities)
- YELLOW: Release and Learning (Release Gate, PMS)

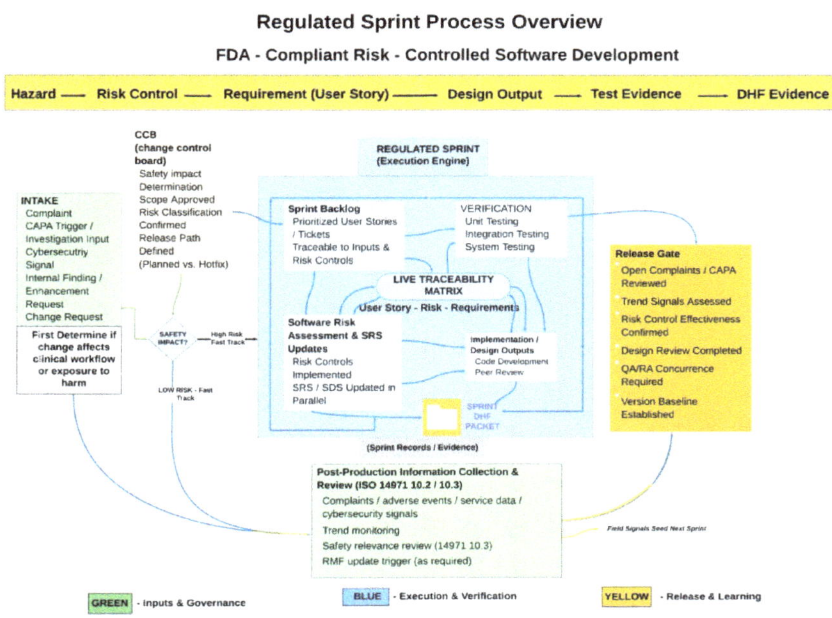

Figure FFF-1: Regulated Sprint Process Overview — Complete FDA and EU MDR Compliant Risk-Controlled Development Framework (Full-Size Reference)

The traceability chain at the top (Hazard → Risk Control → Requirement → Design Output → Test Evidence → DHF Evidence) represents the fundamental principle: every change traces from risk through implementation to verification evidence.

Key insight: Field signals from Post-Market Surveillance seed the next sprint, creating a continuous improvement loop that satisfies both Agile iteration and regulatory post-market requirements.

Appendix GGG — Comprehensive Design Control Tracker

This appendix presents a reference Design Control Tracker that illustrates how to implement the framework concepts described in this book. Organizations implementing this methodology should adapt the tracker structure to their specific quality management system and validate it per their procedures (21 CFR 820.70(i) / ISO 13485 Clause 4.1.6) before production use. The reference tracker demonstrates:

- Dashboard — Real-time release readiness metrics
- Tracker — Full change tracking with SRS/SDS/Risk/Test status
- Bug/Anomaly Resolution — Root cause analysis and corrective action
- Change Request Form — Structured intake with risk decision tree
- Legend — Status definitions and color coding
- User Guide — Implementation instructions
- CCB_Log — Classification decisions with rationale and risk controls
- Verification_Log — Test execution tracking with evidence links
- Sprint_DHF_Packets — Release-ready packet assembly status
- Live_Traceability_Matrix — Bidirectional requirement-to-test traceability
- Risk_Gate_Log — Risk gate decisions and escalations
- Hazard_Register — FMEA/hazard analysis per ISO 14971
- SOUP_Register — Third-party software tracking per IEC 62304
- Sprint_Board — Regulated sprint visual board

Note: Version 28 of the tracker includes 21 integrated sheets providing comprehensive coverage from intake through release, aligned with QMSR's requirement for "linked processes."

GGG.1 Dashboard View

The Dashboard provides at-a-glance release readiness. Key metrics include total changes, release-ready count, blockers, and items in progress. The Status Summary by Stage shows how many items are at each documentation stage (SRS, SDS, Risk, Test).

Figure GGG-1: Design Control Tracker Dashboard — Sprint Release Readiness with Gates

GGG.2 Main Tracker View

The main Tracker sheet is the core working view. Each row represents a change with columns for: Feature/Change, Module, SRS Status, SDS Status, Risk Status, Test Status, Owner, Target Release, Risk Level, Regulatory Impact, Evidence Location, Release Status, Blockers, Next Action, and IEC 62304 Stage.

Color coding makes status instantly visible: Approved (green), In Progress (blue), Under Review (yellow), Needs Update (orange), Draft (gray), Not Started (white), Blocked (red).

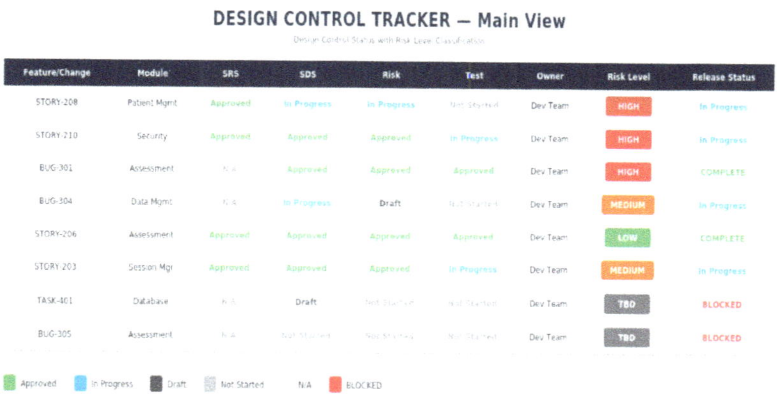

DESIGN CONTROL TRACKER — Main View

Design Control Status with Risk Level Classification

Feature/Change	Module	SRS	SDS	Risk	Test	Owner	Risk Level	Release Status
STORY-208	Patient Mgmt	Approved	In Progress	In Progress	Not Started	Dev Team	HIGH	In Progress
STORY-210	Security	Approved	Approved	Approved	In Progress	Dev Team	HIGH	In Progress
BUG-301	Assessment	N/A	Approved	Approved	Approved	Dev Team	HIGH	COMPLETE
BUG-304	Data Mgmt	N/A	In Progress	Draft	Not Started	Dev Team	MEDIUM	In Progress
STORY-206	Assessment	Approved	Approved	Approved	Approved	Dev Team	LOW	COMPLETE
STORY-203	Session Mgr	Approved	Approved	Approved	In Progress	Dev Team	MEDIUM	In Progress
TASK-401	Database	N/A	Draft	Not Started	Not Started	Dev Team	TBD	BLOCKED
BUG-305	Assessment	N/A	Not Started	Not Started	Not Started	Dev Team	TBD	BLOCKED

Figure GGG-2a: Main Tracker View — Design Control Status with Risk Level Classification

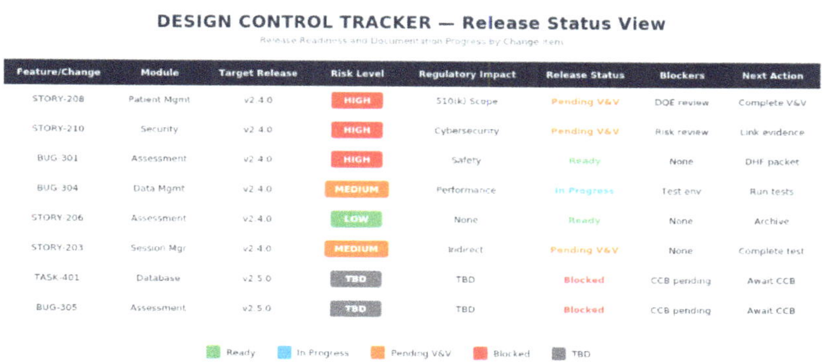

DESIGN CONTROL TRACKER — Release Status View

Release Readiness and Documentation Progress by Change Item

Feature/Change	Module	Target Release	Risk Level	Regulatory Impact	Release Status	Blockers	Next Action
STORY-208	Patient Mgmt	v2.4.0	HIGH	510(k) Scope	Pending V&V	DOE review	Complete V&V
STORY-210	Security	v2.4.0	HIGH	Cybersecurity	Pending V&V	Risk review	Link evidence
BUG-301	Assessment	v2.4.0	HIGH	Safety	Ready	None	DHF packet
BUG-304	Data Mgmt	v2.4.0	MEDIUM	Performance	In Progress	Test env	Run tests
STORY-206	Assessment	v2.4.0	LOW	None	Ready	None	Archive
STORY-203	Session Mgr	v2.4.0	MEDIUM	Indirect	Pending V&V	None	Complete test
TASK-401	Database	v2.5.0	TBD	TBD	Blocked	CCB pending	Await CCB
BUG-305	Assessment	v2.5.0	TBD	TBD	Blocked	CCB pending	Await CCB

Figure GGG-2b: Main Tracker View — Release Status and Documentation Progress

GGG.3 Bug/Anomaly Resolution

The Bug/Anomaly Resolution sheet tracks defects with full root cause analysis. Fields include: Bug ID, Title, Source, Date Reported, Affected Version, Affected Module, Severity, Safety Impact, Risk File Update Required, Linked SRS ID, Linked Risk ID, Root Cause, Corrective Action, Assigned To, Target Version, Regression Test ID, Status, Closure Evidence, Date Closed, and Notes.

This structure ensures bug resolution meets IEC 62304 problem resolution requirements and maintains full traceability to risk and requirements.

CCB LOG — Change Control Board Decisions

Classification decisions with risk rationale and required controls

CCB_ID	Date	Item_ID	Title	Risk Level	Severity	Probability	Decision	Verification Req
CCB-24-004	2024-01-15	STORY-208	Batch patient import	HIGH	Serious	Probable	**Approved**	Full V&V
CCB-24-005	2024-01-15	STORY-210	Multi-factor auth	HIGH	Serious	Remote	**Approved**	Full V&V
CCB-24-006	2024-01-15	BUG-304	Timezone incorrect	MEDIUM	Marginal	Probable	**Approved**	Unit + Integration
CCB-24-007	2024-01-15	STORY-206	Progress indicator	LOW	Negligible	Remote	**Approved**	Unit Test
CCB-24-008	2024-01-16	BUG-301	Score calculation	HIGH	Serious	Frequent	**Approved**	Full V&V + Regr
CCB-24-009	2024-01-16	STORY-203	Session timeout	MEDIUM	Marginal	Occasional	**Approved**	Unit + Integration

Severity (ISO 14971) | Probability of Occurrence | Risk Level = Severity × Probability Matrix

Figure GGG-3: CCB Log — Classification Decisions with P1/P2 Risk Rationale

GGG.4 Change Request Form

The Change Request Form provides a structured intake template. It includes Change Identification, Intended Use Check (four key questions about clinical workflow and harm exposure), Risk Classification with justification, Required Actions (SRS, SDS, Risk File, Regression Testing), and Approvals.

v4 governance update: For HIGH items, the form should capture the preliminary causal chain (Hazard, Hazardous Situation, Harm), P1 basis (data/source supporting exposure probability estimate), and P2 basis (clinical rationale for harm probability given exposure). These fields feed the CCB classification rationale record and the risk file. If not yet present, add to the Change Identification / Impact Assessment section before Safety_Impact.

The form implements the CCB pre-classification process described in Chapter 9, ensuring every change is properly assessed before entering the sprint backlog.

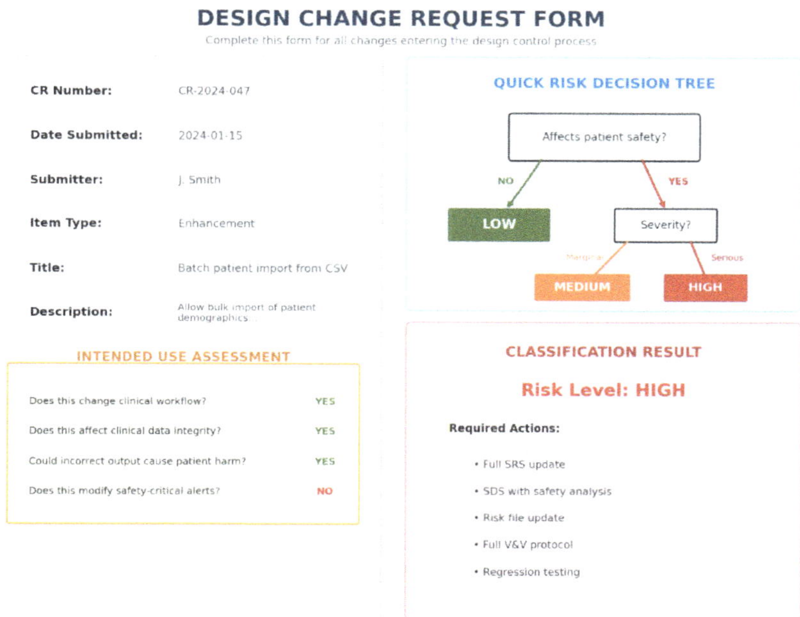

Figure GGG-4: Design Change Request Form — Intake with Quick Risk Decision Tree

GGG.5 Using the Tracker

Implementation guidance:

1. Download the reference tracker
2. Customize the Module dropdown values for your product architecture
3. Adjust the Risk Level criteria to match your organization's risk policy
4. Train team members on status definitions using the Legend sheet
5. Use the Dashboard in sprint reviews and release gate meetings
6. Archive completed sprints by copying to dated backup files

The tracker can be used standalone or as a companion to Jira/Azure DevOps. Many organizations use this tracker as the 'source of truth' view for regulatory evidence while managing day-to-day work in their ALM tool.

Appendix HHH — Book Example Tracker Views

This appendix provides tracker views using the specific examples from this book, including BUG-4521 (Appendix B walkthrough), FEAT-892 (Appendix H HIGH risk feature), and Sprint 47 scenarios. These figures show exactly how the book's examples appear in a configured tracking system.

HHH.1 BUG-4521 Lifecycle Example

This figure shows the complete lifecycle of BUG-4521 from the walkthrough in Appendix B. The timeline shows exactly when classification occurred (before sprint entry) and when evidence was linked (before closure).

LIVE TRACEABILITY MATRIX

Real-time view: Requirement → Design → Hazard → Test → Result

Item_ID	Risk	Req_ID	Req_Status	Design_ID	Hazard_ID	Test_ID	Result
STORY-201	MEDIUM	SRS-201	Approved	SDS-3.2.1	HAZ-012	TC-201-01	PASS
STORY-204	LOW	SRS-204	Approved	SDS-4.1.2	—	TC-204-01	PASS
STORY-205	LOW	SRS-205	Approved	SDS-2.3.1	—	TC-205-01	PASS
BUG-301	HIGH	SRS-042	Approved	SDS-5.1.1	HAZ-008	TC-301-01	PASS
STORY-208	HIGH	SRS-208	In Progress	SDS-6.2.1	HAZ-015	TC-208-01	PENDING
STORY-210	HIGH	SRS-210	Approved	SDS-1.4.2	HAZ-003	TC-210-01	IN PROGRESS
BUG-304	MEDIUM	SRS-119	Draft	SDS-7.1.3	HAZ-021	TC-304-01	NOT STARTED

RESULT: PASS IN PROGRESS PENDING NOT STARTED FAIL

Traceability ensures every requirement links to design, hazard analysis, and verified test results

Figure HHH-1: Live Traceability Matrix — Requirement to Test with Risk Level

HHH.2 FEAT-892 HIGH Risk Feature

This figure shows the FEAT-892 Critical Alert Feature from Appendix H. The epic is classified HIGH because it affects safety-critical alerting. Child stories inherit appropriate classifications based on their specific safety impact.

CCB LOG — FEAT-892 HIGH Risk Feature

Classification decisions for Critical Alert epic and child stories

CCB_ID	Date	Item_ID	Title	Risk Level	Severity	Probability	Decision	Verification Req	Note
CCB-2026-032	2026-01-15	FEAT-892	Critical Alert (Epic)	HIGH	Serious	Probable	**Approved**	Full V&V + Regr	Epic classification drives child story assessment
CCB-2026-033	2026-01-15	FEAT-892-1	Threshold config UI	MEDIUM	Marginal	Probable	**Approved**	Unit + Integ	
CCB-2026-034	2026-01-15	FEAT-892-2	Alert trigger logic	HIGH	Serious	Probable	**Approved**	Full V&V	
CCB-2026-035	2026-01-15	FEAT-892-3	Visual alert indicator	HIGH	Serious	Probable	**Approved**	Full V&V	
CCB-2026-036	2026-01-15	FEAT-892-4	Audio alert tone	HIGH	Serious	Frequent	**Approved**	Full V&V + Regr	
CCB-2026-037	2026-01-15	FEAT-892-5	Alert ack workflow	HIGH	Serious	Occasional	**Approved**	Full V&V	
CCB-2026-038	2026-01-16	FEAT-892-6	Alert history logging	MEDIUM	Marginal	Remote	**Approved**	Unit Test	

Epic-level classification cascades to child stories | Each story assessed independently at CCB

Figure HHH-2: CCB Log — FEAT-892 HIGH Risk Feature Classification

HHH.3 CCB Classification Examples

This figure demonstrates CCB classification decision-making. The color coding makes safety impact immediately visible: HIGH (red), MEDIUM (yellow), LOW (green).

RISK MATRIX

Severity →

Probability ↓	Negligible	Marginal	Serious	Critical
Remote	LOW	LOW	MED	MED
Occasional	LOW	MED	MED	HIGH
Probable	LOW	MED	HIGH	HIGH
Frequent	MED	MED	HIGH	HIGH

Severity (ISO 14971)

CLASSIFICATION EXAMPLES

Item	Severity	Probability	Result
CHG-101: Drug interaction algorithm	Serious	Probable	HIGH
CHG-102: Alert audio frequency	Serious	Probable	HIGH
CHG-103: DB query optimization	Marginal	Occasional	MEDIUM
CHG-104: Dose confirmation dialog	Marginal	Probable	MEDIUM
CHG-105: Copyright year in footer	Negligible	Remote	LOW
CHG-106: Help documentation typo	Negligible	Remote	LOW
CHG-107: Login timeout change	Marginal	Occasional	MEDIUM
CHG-108: Patient photo verification	Marginal	Occasional	MEDIUM

Figure HHH-3: CCB Classification Examples with P1 × P2 Risk Matrix

HHH.4 Sprint DHF Packets

This figure shows Sprint DHF Packets — the release-ready Design History File packet assembly status tracked by sprint. Each row represents a sprint with counts of HIGH, MEDIUM, and LOW items along with verification status, risk file update status, and traceability completion. Sprint 22 and Sprint 23 are fully Approved with complete verification and traceability. Sprint 24 is In Progress, with verification and partial traceability still pending.

SPRINT DHF PACKETS

Release-ready Design History File packets by sprint

Sprint	DHF_Packet_ID	Status	Items	HIGH	MED	LOW	Verified	Risk Upd	Trace
Sprint 22	SP22-DHF-001	Approved	10	1	3	6	YES	YES	COMPLETE
Sprint 23	SP23-DHF-001	Approved	12	2	4	6	YES	YES	COMPLETE
Sprint 24	SP24-DHF-001	In Progress	18	2	6	10	NO	YES	PARTIAL

Each Sprint DHF Packet = smallest auditable unit containing all evidence for that sprint

Packet approved only when: All items verified + Risk file updated + Traceability complete

Figure HHH-4: Sprint DHF Packets — Release-Ready Evidence Package

HHH.5 Traceability Matrix

This figure demonstrates bidirectional traceability: Requirement → Design → Hazard → Test → Result. The color-coded Result column shows verification status at a glance.

LIVE TRACEABILITY MATRIX — Bidirectional Links

Requirement → Design → Hazard → Test → Result

Item_ID	Risk	Req_ID	Req_Status	Design_ID	Hazard_ID	Test_ID	Result
FEAT-892-1	MEDIUM	SRS-9.0	Approved	SDS-9.0.1	HA-0044	TC-0111	PASS
FEAT-892-2	HIGH	SRS-9.1	Approved	SDS-9.1.1	HA-0044	TC-0112	PASS
FEAT-892-3	HIGH	SRS-9.2	Approved	SDS-9.2.1	HA-0045	TC-0113	PASS
FEAT-892-4	HIGH	SRS-9.3	Approved	SDS-9.3.1	HA-0046	TC-0114	PENDING
FEAT-892-5	HIGH	SRS-9.4	Approved	SDS-9.4.1	HA-0047	TC-0115	IN PROGRESS
FEAT-892-6	MEDIUM	SRS-9.5	Approved	SDS-9.5.1	HA-0049	TC-0116	PASS
BUG-4521	MEDIUM	SRS-4.2	Approved	SDS-4.2.1	HA-0052	TC-0089	PASS
CHG-104	MEDIUM	SRS-5.1	Approved	SDS-5.1.1	HA-0015	TC-0078	PASS

RESULT: PASS PENDING IN PROGRESS NOT STARTED FAIL

Figure HHH-5: Live Traceability Matrix — Bidirectional Links with Status

HHH.6 Release Gate Dashboard

This figure shows the Release Gate Dashboard. The evidence status matrix shows completion at a glance. The GATE STATUS: HOLD demonstrates that the framework prevents release when evidence is incomplete.

Figure HHH-6: Release Gate Dashboard — Sprint Readiness with Gate Status

Appendix III — Essential SOP Architecture

This appendix defines the minimum set of Standard Operating Procedures required to support the framework. Organizations may consolidate or expand these based on size and complexity, but the functional coverage must exist. Each SOP should reference specific framework elements and regulatory requirements.

Required Procedures

SOP-001: Design and Development Control

Governs the overall design control process per 21 CFR 820.30 and ISO 13485 Clause 7.3. References: Chapters 8-12 (Regulated Sprint, Risk Gate, Parallel Execution, Tracker, Tools). Defines planning, input, output, review, verification, validation, transfer, and change control requirements.

SOP-002: Risk Management

Establishes risk management process per ISO 14971:2019. References: Chapter 9 (Risk Gate), Appendix P (Risk File Structure). Covers risk analysis, evaluation, control, residual risk evaluation, and production/post-production monitoring. Must include P1/P2 probability definitions and acceptability criteria.

SOP-003: Software Development Lifecycle

Defines software lifecycle per IEC 62304:2006+A1:2015. References: Appendix AAA (IEC 62304 Integration), Appendix BBB (SDP Template). Addresses software safety classification, development planning, requirements, architecture, implementation, integration, testing, release, and maintenance.

SOP-004: Change Control Board Operations

Governs CCB structure, membership, meeting cadence, classification criteria, and documentation requirements. References: Chapter 9 (Risk Gate), Appendix T (CCB Classification Scenarios). Must include the Universal Question, classification definitions (HIGH/MEDIUM/LOW), and escalation criteria.

SOP-005: Verification and Validation

Defines V&V requirements per 21 CFR 820.30(f-g) and IEC 62304 Clause 5.5-5.7. References: Chapter 15 (Release Gates), Appendix Q (Verification

Testing), Appendix R (Validation vs. Verification). Includes protocol requirements, acceptance criteria, and evidence documentation.

SOP-006: Document and Record Control

Establishes document control per 21 CFR 820.40 and ISO 13485 Clause 4.2. References: Appendix NN (Documentation Management). Covers DHF structure, version control, review/approval workflows, and retention requirements. Must address electronic records per 21 CFR Part 11.

SOP-007: Corrective and Preventive Action (CAPA)

Defines CAPA process per 21 CFR 820.100 and ISO 13485 Clause 8.5. References: Appendix BB (CAPA Integration). Covers problem identification, investigation, root cause analysis, corrective action, preventive action, and effectiveness verification.

SOP-008: Complaint Handling

Governs complaint intake, evaluation, investigation, and trending per 21 CFR 820.198. References: Appendix J (Complaint to Resolution Walkthrough), Chapter 14 (Post-Market Surveillance). Includes MDR/vigilance reporting criteria.

SOP-009: Supplier and Third-Party Control

Establishes supplier qualification and monitoring per 21 CFR 820.50 and ISO 13485 Clause 7.4. References: Appendix CC (Supplier Control). Addresses SOUP/OTS evaluation, supplier audits, and incoming quality requirements.

SOP-010: Training and Competency

Defines training requirements per 21 CFR 820.25 and ISO 13485 Clause 6.2. References: Appendix W (Training Requirements). Includes role-based training matrices, competency assessment methods, and training record maintenance.

Implementation Note: These ten procedures represent the minimum functional coverage. Smaller organizations may combine SOPs (e.g., merge SOP-001 and SOP-003 into a single Design Control procedure). Larger organizations may decompose them further. The key is complete regulatory coverage with clear cross-references to framework elements. Each SOP should explicitly state which framework chapters and appendices it implements.

Appendix JJJ — IEC 62304 Edition 2 Transition Advisory

IEC 62304 Edition 2 is currently in draft (public comment closed February 2025, projected publication August 2026). This appendix provides forward-looking guidance for organizations preparing for the transition while maintaining compliance with the current enforceable standard, IEC 62304:2006+A1:2015 (Edition 1.1).

JJJ.1 Key Anticipated Changes

Scope Expansion: Edition 2 broadens scope from 'medical device software' to 'health software,' aligning with IEC 82304-1 (health software products) and IEC 81001-5-1 (health IT security). Organizations developing health software beyond traditional medical devices should monitor this expansion.

Classification Restructuring: The current three-class model (Class A/B/C) is expected to transition to a two-level rigor model (Level I/Level II). Level I corresponds approximately to current Class A, while Level II encompasses current Classes B and C. This simplification eliminates the B/C differentiation that many organizations find difficult to apply consistently. The framework's CCB classification (HIGH/MEDIUM/LOW) maps naturally to both models: HIGH maps to Class C/Level II, MEDIUM maps to Class B/Level II, LOW maps to Class A/Level I.

Legacy Software Pathway: The current normative Clause 4.4 legacy software compliance pathway is anticipated to move to an informative annex. Organizations with legacy software should complete Clause 4.4 gap analyses under Edition 1.1 before Edition 2 adoption, as the normative obligation provides stronger regulatory defensibility than informative guidance.

AI/ML Process Planning: Edition 2 is expected to include formalized requirements for AI/ML lifecycle governance, aligning with the PCCP framework described in Appendix KK and the ENQUAL Knowledge System Section LII.

Probability Language: The foundational rule 'probability of software failure contributing to a hazardous situation is assumed to be 1' is expected to transition to 'worst-case scenario assumption' language. The intent is unchanged — only external risk control measures (not software reliability claims) can reduce safety classification — but the simplified language should reduce industry confusion.

JJJ.2 Transition Planning for Framework Users

Phase 1 — Now through Pre-Publication: Continue full compliance with IEC 62304:2006+A1:2015. Complete pending legacy software assessments under the normative Clause 4.4. Document software safety classifications in a format that maps to both three-class (A/B/C) and two-level (I/II) models. The framework's CCB classification records already support this dual mapping through the Safety_Impact field.

Phase 2 — Publication through Adoption Period: Conduct gap analysis between current framework governance and Edition 2 requirements. Identify engagements affected by scope expansion (health software vs. medical device software). Update the clause-by-class applicability matrix (Appendix AAA Table AAA-5) to reflect the new Level I/II structure. Monitor regulatory body adoption timelines — EU MDR/IVDR harmonization and FDA recognition will drive actual compliance obligations.

Phase 3 — Regulatory Adoption: Transition active engagements per regulatory timeline when Edition 2 is harmonized under EU MDR/IVDR or recognized by FDA. A three-year transition period is recommended for organizations with complex software portfolios.

REGULATORY ANCHOR: Previous IEC 62304 revision attempts have not reached publication. Execution risk exists for Edition 2. All framework governance should be built against the current enforceable standard (Edition 1.1) with forward compatibility provisions.

KEY INSIGHT: The framework's risk-based CCB classification already provides a natural bridge between Edition 1.1 and Edition 2. Organizations following this framework will have minimal transition effort because the underlying principle — proportional rigor based on safety impact — remains constant across both editions.

Appendix KKK — AAMI HE75:2025 Updated Human Factors Guidance

AAMI HE75:2025 supersedes HE75:2009/(R)2013/(R)2018 as the AAMI American National Standard for human factors engineering design of medical devices. Approved 30 April 2025 by AAMI and 30 October 2025 by ANSI, this edition provides updated guidance that complements IEC 62366-1 requirements covered in Appendices WW through ZZ.

KKK.1 Key Updates from HE75:2009 to HE75:2025

The 2025 edition reflects evolution in medical device technology and human factors practice since 2009. For SaMD teams using this framework, the most relevant updates include expanded coverage of software user interfaces (Section 13), updated usability testing guidance (Section 9) incorporating remote and hybrid evaluation methods, enhanced alarm management principles (Section 11) aligned with IEC 60601-1-8 alarm hazard categories, and new considerations for AI-assisted clinical decision support interfaces.

Section 5 (Managing the Risk of Use Error) provides the conceptual framework linking use error to risk management per ISO 14971. This section reinforces the framework's core principle established in Chapter 6: user behavior appears in the hazardous situation, never as the hazard itself. HE75:2025 explicitly states that use errors are consequences of device design characteristics, not independent hazard sources.

Section 9 (Usability Testing) provides updated design principles for formative and summative evaluation that align with the framework's sprint-based usability integration described in Chapter 13 and Appendix YY. HE75:2025 guidance on remote usability evaluation is particularly relevant for SaMD teams conducting formative testing during sprints.

Section 13 (Software User Interfaces) addresses SaMD-specific design considerations including information hierarchy for clinical displays, clinical decision support presentation, and data visualization for risk communication. These principles should inform SRS development for user-facing features classified MEDIUM or HIGH at CCB.

KKK.2 Framework Integration Points

Sprint-Level Application: During parallel execution (Chapter 10), DQE should reference HE75:2025 Section 5 principles when updating risk files for user-facing changes. Formative usability activities during sprints should follow Section 9 guidance for evaluation planning and documentation. Results enter Intake like any other design input per the standard framework process.

CCB Classification: Changes affecting user interface design should be assessed against HE75:2025 design principles during CCB classification. Changes to clinical displays, alert/alarm presentation, or navigation affecting clinical workflow should trigger MEDIUM or HIGH classification based on the universal question: 'Does this affect clinical workflow or patient safety?'

Usability Engineering File: The UEF described in Appendix WW should reference HE75:2025 as the design guidance standard supporting IEC 62366-1 compliance. Basic human skills and abilities (Section 6), anthropometry and biomechanics (Section 7), and UI design rationale (Sections 18-21) documented per HE75:2025 provide the evidence base for usability engineering decisions.

REGULATORY ANCHOR: AAMI HE75:2025 is a guidance standard providing the 'how-to' for achieving IEC 62366-1 usability engineering requirements. While IEC 62366-1 defines what must be done (normative requirements), HE75 describes how to do it effectively. Both standards are recognized by FDA and accepted as demonstrating good human factors engineering practice. The 2025 edition should be referenced in new usability engineering plans; existing plans referencing HE75:2009 should be updated at the next planned revision.

Appendix LLL — EU MDR/IVDR Comprehensive Framework Alignment

Appendix I provided initial EU MDR mapping. This appendix provides expanded coverage addressing EU MDR 2017/745 and EU IVDR 2017/746 requirements that intersect with the regulated sprint framework, including post-market surveillance obligations and Notified Body audit expectations.

LLL.1 EU MDR Post-Market Surveillance Framework (Articles 83-86)

EU MDR establishes PMS obligations that go significantly beyond FDA complaint handling requirements. The regulated sprint framework's PMS loop (Chapter 14) must be extended to address these requirements.

Article 83 (PMS System): Manufacturers must plan, establish, document, implement, maintain, and update a PMS system for each device. The PMS system must be suited to actively collecting and using relevant data for updating the benefit-risk determination, design and manufacturing information, IFU, and risk management. The framework's PMS Manager role (Chapter 7) owns this system. Field signals entering Intake implement the 'active collection' requirement — the framework does not wait passively for complaints but actively monitors device performance in the field.

Article 84 (PMS Plan): Requires a documented PMS plan per device covering: systematic procedures for collecting and evaluating field data, methods for managing reported incidents and field safety corrective actions, methods for tracing and identifying devices needing corrective action, a PMCF plan if applicable, and reference to clinical evaluation and PMCF (Article 61). The PMS Manager maintains this plan with cross-references to the framework's Intake, CCB, and CAPA integration processes.

Article 85 (PMS Report — Class I): For Class I devices, manufacturers produce a PMS report summarizing results and conclusions of PMS data analysis, rationale for preventive and corrective actions, and updates to

Technical Documentation. This report is updated when necessary and made available to competent authorities upon request.

Article 86 (PSUR — Class IIa/IIb/III): For higher-class devices, manufacturers produce a Periodic Safety Update Report (PSUR) which sets out the conclusions of the benefit-risk determination, main findings of PMCF, and the volume and category of all detected incidents and field safety corrective actions. PSURs must be updated at least annually for Class IIb/III and at least every two years for Class IIa devices. PSURs for Class III devices are submitted to the Notified Body.

LLL.2 EU IVDR 2017/746 Specific Considerations

IVD software increasingly overlaps with SaMD. EU IVDR introduces additional requirements beyond EU MDR for in vitro diagnostic devices.

Classification Rules: IVDR classification rules (Annex VIII) differ significantly from MDR Annex VIII. IVD software that performs analysis functions or provides diagnostic results typically falls into Class C or D, requiring Notified Body involvement. The framework's CCB classification must account for IVDR classification rules when the device falls under IVDR scope.

Performance Evaluation: IVDR requires performance evaluation (Articles 56-58) in place of clinical evaluation. For IVD software, this includes analytical performance (sensitivity, specificity, accuracy, precision) and clinical performance (clinical sensitivity, clinical specificity, positive/negative predictive values). These performance characteristics become design inputs documented in the SRS and verified through the framework's verification process described in Chapter 15 and Appendix Q.

Common Specifications: IVDR empowers the European Commission to adopt common specifications (Article 9) for device performance, labeling, PMS, and clinical follow-up. Teams developing IVD software under the regulated sprint framework should monitor common specifications adoption as these create mandatory compliance requirements that may affect design inputs and verification criteria.

LLL.3 Notified Body Audit Alignment

EU MDR/IVDR Notified Body audits differ from FDA inspections in structure and emphasis. The framework prepares for both through comprehensive evidence production, but teams should understand key differences.

Technical Documentation Review: Notified Bodies perform detailed review of Technical Documentation (Annex II/III) as part of conformity assessment. The Sprint DHF Packets described in Chapter 12 and Appendix M compile directly into Technical Documentation sections. Specifically: Sprint DHF evidence populates Annex II Section 3 (Design and Manufacturing Information), risk files populate Annex II Section 4 (GSPR Compliance), and verification evidence populates Annex II Section 6 (Product Verification and Validation).

General Safety and Performance Requirements (GSPR): EU MDR Annex I and IVDR Annex I define GSPRs that must be met. The framework's risk management process (ISO 14971 integration described in Chapter 6 and the causal chain methodology) directly addresses GSPR compliance. Each hazard in the risk file should trace to the specific GSPR it addresses. DQE should maintain this mapping as part of the risk file governance described in Appendix P.

REGULATORY ANCHOR: EU MDR Article 10(9) requires manufacturers to implement and keep up to date a QMS. ISO 13485:2016 certification satisfies this requirement and aligns with FDA QMSR (Appendix TT). The framework's integration of both FDA and EU requirements through a single process — the regulated sprint — enables dual-market compliance without maintaining parallel quality systems.

Appendix MMM — Design Control Deliverables by Phase (Risk-Scaled)

This appendix defines the minimum design control deliverables required at each lifecycle phase, scaled by risk level. This provides a quick-reference for DQE and engineering teams to verify completeness at each phase gate without referencing multiple standards. The deliverables integrate requirements from 21 CFR 820.30, ISO 13485:2016 Clause 7.3, IEC 62304 Clauses 5-9, IEC 62366-1, and ISO 14971.

MMM.1 Planning Phase Deliverables

All risk levels require: Software Development Plan (IEC 62304 Clause 5.1) defining lifecycle model, tools, methods, and standards; Risk Management Plan (ISO 14971 Clause 4.5) defining scope, risk acceptability criteria, and verification planning; Quality Plan or reference to QMS procedures establishing design control requirements for the project.

HIGH risk items additionally require: User Interface Evaluation Plan (IEC 62366-1 Clause 5.7) defining formative and summative evaluation approach; Clinical Evaluation Plan (EU MDR Article 61) if applicable; PCCP if AI/ML components are included; Cybersecurity Plan per FDA 2023 Cybersecurity Guidance if network-connected.

MMM.2 Design Input Phase Deliverables

All risk levels require: Software Requirements Specification (SRS) with acceptance criteria traceable to user needs; Intended Use/Intended Purpose statement (FDA/EU MDR); Software safety classification determination (IEC 62304 Clause 4.3) documented with rationale.

MEDIUM and HIGH risk items additionally require: Use Specification (IEC 62366-1 Clause 5.1) defining intended users, use environments, and user interface; Hazard Analysis with initial hazard identification (ISO 14971 Clause 5); Regulatory classification determination with predicate/equivalence analysis.

HIGH risk items additionally require: Clinical Performance requirements (IVDR Articles 56-58 if IVD); Interface specifications for interoperability (if applicable); Training data requirements and bias analysis (if AI/ML).

MMM.3 Design Output Phase Deliverables

All risk levels require: Software Architecture Description (IEC 62304 Clause 5.3) appropriate to safety class — Class C requires documented architecture showing software items, external interfaces, and SOUP components; Class A may document architecture at a system level; Software Design Specification (SDS) for MEDIUM and HIGH items.

MEDIUM and HIGH risk items additionally require: Risk Control implementation documentation (ISO 14971 Clause 7.1) mapping controls to specific design outputs; SOUP evaluation records (IEC 62304 Clauses 5.3.3/5.3.4) documenting functional requirements, performance requirements, and known anomaly review for each SOUP component; Traceability from design outputs to design inputs.

MMM.4 Verification Phase Deliverables

All risk levels require: Unit test results (IEC 62304 Clause 5.5.5 — required for Class C, recommended for Class B, optional for Class A); Integration test results (IEC 62304 Clause 5.6); System test results (IEC 62304 Clause 5.7 — required for all classes per Amendment 1); DQE verification approval (for MEDIUM and HIGH items).

MEDIUM and HIGH risk items additionally require: Risk control verification evidence (ISO 14971 Clause 7.3) demonstrating each control is implemented and effective; Requirements-based test coverage analysis demonstrating every SRS requirement has verification evidence.

HIGH risk items additionally require: DQE witnessed test execution for safety-critical test cases; RA review confirmation for regulatory submission impact; Design review records (21 CFR 820.30(e)) with cross-functional participation.

MMM.5 Validation Phase Deliverables

Validation activities typically span beyond individual sprints and occur in the Validation Sprint described in Chapter 15.

All risk levels where validation applies require: Validation protocol defining acceptance criteria traceable to user needs and intended use; Validation execution results with documented actual outcomes; Validation report summarizing results and conclusions.

MEDIUM and HIGH risk items additionally require: Summative usability evaluation results (IEC 62366-1 Clause 5.9) with representative users; Clinical evaluation report or performance evaluation report (EU MDR/IVDR); Residual risk evaluation (ISO 14971 Clause 8) documenting overall residual risk acceptability.

MMM.6 Transfer and Release Phase Deliverables

All risk levels require: Release notes documenting version, changes, and known issues; Build verification confirming deployed binary matches verified configuration; SBOM (Software Bill of Materials) current for the release version; Sprint DHF Packet(s) compiled into Release DHF.

MEDIUM and HIGH risk items additionally require: DQE release approval (MEDIUM) or DQE + RA dual release approval (HIGH); Risk Management Report (ISO 14971 Clause 9) current for this release; Complete bidirectional traceability matrix (requirement to test to risk to design).

For EU market release additionally require: Declaration of Conformity; Technical Documentation per Annex II structure; GSPR compliance matrix; UDI assignment and EUDAMED registration.

KEY INSIGHT: These deliverables are not separate from sprint work — they are the natural outputs of the regulated sprint process. When parallel execution (Chapter 10) operates correctly, these deliverables exist because the work produced them, not because someone created them after the fact. The DQE confirms completeness; they do not create the deliverables retroactively.

Appendix NNN — Consolidated Governance Template Library

This appendix consolidates all governance templates from the main text into a single copy-ready reference. Every template is written in strict ISO 14971 form: Hazard → Hazardous Situation → Harm, with P1 and P2 addressed. Templates are inspection-ready: use them verbatim or adapt field labels to your quality management system. The governance rules they enforce remain non-negotiable.

Each template block is color-coded by function. The key below applies throughout this appendix.

▌	Navy Blue (▌▌)	**Administrative / Record Header**	ID, dates, signatures, and decisions. Who owns this record and when was it created.
✓	Forest Green (✓)	**Technical Analysis Required**	Causal chain, probability, and benefit-risk sections. Substantive clinical/technical work is required here — not just administrative completion.
▌	Dark Navy (▌)	**Permanent Reference Standard**	ISO 14971 causal chain quick-reference. Fixed rules that do not change with the record — a standing reference, not a fillable field.
⚠	Red (⚠)	**Non-Negotiable Rule**	Rules whose violation constitutes an ISO 14971 compliance failure. Stop and read — these are not guidelines.
►	Orange (►)	**Decision Point / Routing Rule**	Classification routing and Safety_Impact rigor-level assignment. Use this block to determine which process path applies.

NNN.1 CCB Classification Rationale Record

Required for MEDIUM and HIGH. Recommended for all. This record is the primary inspection artifact for the CCB classification decision. It must be created at the time of the CCB meeting, not reconstructed afterward. The completed record is stored in the tracker (CCB_Reference field) and the DHF.

CCB CLASSIFICATION RATIONALE RECORD

Item ID: ________________ CCB Date: _________________ DQE: ______________
Item Description (what the change is): _____________________________________

✓ CAUSAL CHAIN ANALYSIS

Hazard (device characteristic / information for safety):
What characteristic of the device creates the potential for harm?
NOT: user behavior, defect labels, or process failures.
Entry: ____

Hazardous Situation (exposure circumstance / incorrect interaction):
When / how is the patient or user exposed to the hazard?
Use error belongs HERE, not at the Hazard level.
Entry: ____
Harm (clinical consequence):
What is the injury or damage outcome if the hazardous situation leads to harm?
Most likely harm: ___________________________________
Worst plausible harm (clinically realistic maximum): _____________
Severity level selected: __________ Rationale: _________________

✓ PROBABILITY ANALYSIS

P1 basis (exposure probability — probability hazardous situation occurs):
Data source: [ ] Usability study [ ] Literature [ ] Field data [ ] Judgment
Source citation / rationale: __________________________ Estimated P1: __________

P2 basis (harm probability given exposure):
Clinical rationale for harm probability if hazardous situation occurs:
Source citation / rationale: ________________________________
Estimated P2: __________
Impacted causal chain ID(s) / Risk_Link(s): [Specific HA-IDs in the risk file — e.g., HA-0023.
Inspector question: show me where this lives in the risk file; ______________________
Affected risk control(s) (existing controls modified or new controls required):

CLASSIFICATION DECISION

Safety_Impact: [ ] HIGH [ ] MEDIUM [ ] LOW Why this classification (rigor rationale): _________

GOVERNANCE REMINDER: Safety_Impact routes rigor level — it is not the ISO 14971 risk estimate. The complete ISO 14971 risk estimate (severity × P1 × P2) is in the risk management file. This record is a pointer to that analysis.

DQE Signature: __________________________ Date: __________________
Risk File Updated: [ ] Yes (HA-ID: _______) [ ] Not required (LOW, rationale: _______)

NNN.2 Jira Ticket Required Governance Fields

Five fields transform a Jira ticket into a design history record. All five must be populated before an item moves to Done for MEDIUM and HIGH items. Inspectors will query these fields to reconstruct the design history.

Field	Content Required	When Set	Inspector Expects
Safety_Impact	HIGH / MEDIUM / LOW Set by DQE at CCB	Before sprint entry	CCB record supporting this classification
CCB_Reference	CCB meeting ID + date (e.g., CCB-2024-0892)	At CCB	Matching CCB log entry with rationale
SRS_Link	SRS section ID(s) addressed (e.g., SRS-4.2.1)	During sprint	SRS section timestamp ≤ sprint dates
Risk_Link	Causal chain ID(s) in risk file (e.g., HA-0023, HA-0045) NOT a folder or document name	During sprint (DQE sets)	Risk file entry at that ID, with P1/P2 and residual risk status
Verification_Status	PENDING → APPROVED DQE approval with timestamp	At verification complete	Approval timestamp ≤ sprint close date

NNN.3 Sprint-Level Definition of Done Checklist

Three sign-offs are required at sprint level before any item is marked Done. This checklist is run at sprint close for every MEDIUM and HIGH item. LOW items use a streamlined version (marked below).

SPRINT DEFINITION OF DONE — ITEM-LEVEL CHECKLIST

Item ID: ____________ Sprint: ___________ Classification: _____________

ENGINEERING SIGN-OFF

☐ Implementation complete and code reviewed

☐ Unit tests passing

☐ SRS_Link populated with correct section ID(s)

☐ Risk_Link populated with specific causal chain ID(s) [MEDIUM/HIGH only]

☐ Verification test cases executed and results linked [MEDIUM/HIGH only]

Engineering Lead: ___________________________ Date: ________________

PRODUCT SIGN-OFF

☐ SRS updated or confirmed current (no change required)

☐ Acceptance criteria met per SRS

☐ No undocumented scope changes

Product Owner: ___________________________ Date: ________________

DQE SIGN-OFF

☐ Risk file reviewed and current (HA-ID confirmed) [MEDIUM/HIGH only]

☐ P1 and P2 estimates updated if new evidence available [MEDIUM/HIGH only]

☐ Risk controls confirmed adequate and verified [MEDIUM/HIGH only]

☐ Verification evidence reviewed and Verification_Status = APPROVED

☐ Causal chain discipline confirmed: no defect labeled as hazard; no user error at Hazard level; Hazardous Situation documented

DQE: ___________________________ Date: ________________

NNN.4 Release Gate Checklist

The release gate is confirmatory, not evaluative. If sprints execute correctly, this checklist confirms existing evidence. Surprises at the release gate indicate a sprint-level breakdown, not a release-level problem. For HIGH items, the benefit-risk conclusion must be documented in writing before dual sign-off.

Note on risk controls: Training-only controls are not sufficient as the primary risk control for MED or HIGH items unless justified against the inherent safety/protective measures hierarchy (ISO 14971 Clause 7.1). A MED or HIGH item whose risk file lists only training or labeling as its primary control is a gate-blocking finding until the hierarchy evaluation is documented.

Check	Status	Evidence Reference
All items have Safety_Impact classification (no unclassified items)	☐ Pass ☐ Fail	Tracker export
All HIGH items: DQE + RA dual approval at sprint level	☐ Pass ☐ Fail ☐ N/A	Verification_Status + approval records
All MEDIUM items: DQE approval	☐ Pass ☐ Fail ☐ N/A	Verification_Status records
All evidence links populated (SRS_Link, Risk_Link, Verification_Status)	☐ Pass ☐ Fail	Tracker field audit
Traceability report generated and reviewed	☐ Pass ☐ Fail	Traceability matrix export
HIGH items: benefit-risk conclusion documented (most likely harm, worst plausible harm, P1 evidence, P2 evidence)	☐ Pass ☐ Fail ☐ N/A	Release record / DHF
Risk file current: all causal chains reflect current design	☐ Pass ☐ Fail	RMF review date
Post-market signals: all signals mapped to causal chain IDs in risk file (no parallel taxonomy; new chains added where required)	☐ Pass ☐ Fail ☐ N/A	PMS review log
No open CAPAs affecting release scope	☐ Pass ☐ Fail ☐ N/A	CAPA log
No unresolved complaints affecting release scope	☐ Pass ☐ Fail ☐ N/A	Complaint log

RELEASE DECISION

Program Manager: ______________________________ Date: ________________

DQE: ______________________________ Date: ________________

RA (required if any HIGH items): ______________________________ Date: ________________

Decision: [ ] Release Approved [ ] Release Blocked — Reason: ____________________

NNN.5 HIGH-Item Benefit-Risk Conclusion Record

Required for every HIGH-classified item at the release gate. This record documents the benefit-risk determination that makes the release decision inspection-defensible. It is not a substitute for the ISO 14971 risk file analysis — it is a release-level summary that references the risk file. The release gate approval signature is insufficient without this record.

BENEFIT-RISK CONCLUSION RECORD — HIGH ITEM

Item ID: ____________ Release: _____________ Risk File Ref: _____________

Causal Chain Summary (from risk file):

Hazard: ____

Hazardous Situation: ____________________________________

Most likely harm: _____________________________________

Worst plausible harm (clinically realistic maximum): __________________

Severity level: ___________ Rationale: ______________________

P1 (exposure probability):

Residual P1 after controls: ___________

Evidence basis: ____

P2 (harm probability given exposure):

Residual P2 after controls: ___________

Evidence basis: ____

Risk controls verified:

[ ] Verification evidence confirms control effectiveness

Evidence reference: _______________________________________

✓ BENEFIT-RISK DETERMINATION

Benefits of this feature/change: _____________________________

Residual risk summary: ____________________________________

Conclusion: The benefit of [describe benefit] outweighs the residual risk of [describe residual risk] because [clinical/technical rationale].

Residual risk is: [ ] Acceptable [ ] Acceptable given benefits [ ] Not acceptable

DQE: ___________________________ Date: ________________

RA: ___________________________ Date: ________________

NNN.6 Post-Market Signal Normalization Record

Every post-market signal — complaint, MDR, field correction, surveillance trend — must be normalized into the existing causal chain structure before it can be closed. This record prevents PMS from becoming a parallel taxonomy disconnected from the risk management file. If no existing chain covers the signal, a new chain must be added to the risk file as part of signal closure.

PMS SIGNAL NORMALIZATION RECORD

Signal ID: _______ Date Received: ________ Source: ____________

Signal description (verbatim from complaint/MDR/report): ______________________

✓ CAUSAL CHAIN MAPPING

Step 1 — Map to causal chain:

Does this signal map to an existing chain? [ ] Yes — Chain ID: ____

[ ] No (new chain required)

Hazard (device characteristic implicated): ________________________

Hazardous Situation (exposure / use error described in signal): ________

Harm (clinical outcome reported or potential): ____________________

Step 2 — P1 impact assessment:

Does this signal change the P1 (exposure probability) estimate?

[ ] Yes — Updated P1: _______ Basis: ________________________

[ ] No — Rationale: _______________________________________

Step 3 — P2 / severity impact:

Does this signal change the P2 estimate or reveal underestimated severity?

[ ] Yes — Updated P2/severity: _______ Basis: _____________

[ ] No — Rationale:

____

Step 4 — Resulting action:

[ ] No action required (P1, P2, severity unchanged, residual risk acceptable)

[ ] Risk file update only (update P1/P2 estimates; no control change needed)

[ ] CAPA required (risk control modification or addition needed)

[ ] Design change required (enters Intake → CCB)

[ ] New causal chain added to risk file (Chain ID assigned: _______)

[ ] Regulatory reporting required (MDR/FSCA)

Risk file updated: [ ] Yes (HA-ID: _______) [ ] Not required (rationale above)

Signal closed by: ___________________ Date: ________________

NNN.7 Causal Chain Quick-Reference Card

Post this card wherever risk files are written and CCB classification decisions are made. The four rules and three-level structure prevent the most common ISO 14971 compliance failures seen in FDA inspections.

ISO 14971 CAUSAL CHAIN — QUICK REFERENCE

▶ **HAZARD = device characteristic / information for safety**

Never: 'user error' • Never: 'software defect' • Never: 'design defect'

Ask: What property of the device creates the potential for harm?

▶ **HAZARDOUS SITUATION = exposure circumstance**

Use error / incorrect interaction belongs HERE.

Ask: When/how is someone exposed to the hazard?

▶ **HARM = clinical / injury outcome**

Ask: What injury or damage occurs if the hazardous situation leads to harm?

Document BOTH most likely harm AND worst plausible harm.

⚠ FOUR NON-NEGOTIABLE RULES

Rule 1: Never call a defect a hazard.

Rule 2: Never call user error a hazard.

Rule 3: Never skip the Hazardous Situation.

Rule 4: Always separate P1 from P2.

P1 = probability of exposure (URRA / IEC 62366-1 refines this)

P2 = probability of harm given exposure (ISO 14971 governs this)

URRA does NOT replace ISO 14971 risk analysis.

▸ SAFETY_IMPACT ROUTING RULE

Safety_Impact routes RIGOR level — it is NOT the ISO 14971 risk estimate.

Safety_Impact routes rigor; ISO 14971 evaluates risk in the risk file using

Hazard → Hazardous Situation → Harm with P1/P2

REFERENCES AND STANDARDS

Regulatory Standards

21 CFR Part 820 — Quality System Regulation (QSR), U.S. Food and Drug Administration

21 CFR Part 11 — Electronic Records; Electronic Signatures

FDA Quality Management System Regulation (QMSR), Final Rule 2024

EU Regulation 2017/745 — Medical Device Regulation (EU MDR)

EU Regulation 2017/746 — In Vitro Diagnostic Regulation (EU IVDR)

International Standards

ISO 13485:2016 — Medical devices — Quality management systems — Requirements for regulatory purposes

ISO 14971:2019 — Medical devices — Application of risk management to medical devices

IEC 62304:2006+A1:2015 (Edition 1.1) — Medical device software — Software life cycle processes

IEC 62366-1:2015/AMD1:2020 — Medical devices — Application of usability engineering to medical devices

ISO/IEC 27001 — Information security management systems

IEC 81001-5-1:2021 — Health software and health IT systems safety, effectiveness and security

AAMI HE75:2025 — Human factors engineering — Design of medical devices (supersedes HE75:2009/(R)2013/(R)2018)

IEC 82304-1:2016 — Health software — Part 1: General requirements for product safety

FDA Compliance Program CP 7382.850 — Quality Management System Inspection (effective February 2, 2026)

AAMI CR34971:2023 — Application of ISO 14971 to Artificial Intelligence and Machine Learning

FDA Guidance Documents

FDA Guidance: Software as a Medical Device (SaMD): Clinical Evaluation (2017)

FDA Guidance: Content of Premarket Submissions for Device Software Functions (2023)

FDA Guidance: Cybersecurity in Medical Devices: Quality System Considerations (2023)

FDA Guidance: Clinical Decision Support Software (2022)

FDA Guidance: Predetermined Change Control Plans for AI/ML-Enabled Devices (2023)

FDA Guidance: Computer Software Assurance for Production and Quality Management System Software (February 3, 2026)

Technical Information Reports

AAMI TIR45:2023 — Guidance on the use of agile practices in the development of medical device software

AAMI TIR57:2016/(R)2019 — Principles for medical device security — Risk management

AAMI HE75:2009/(R)2018 — Human factors engineering — Design of medical devices (superseded by HE75:2025)

AAMI TIR59:2017 — Cybersecurity risk management in medical device systems

IMDRF/SaMD WG/N41 — Software as a Medical Device: Clinical Evaluation

Appendix OOO: Large-Format Figure and Table Reference

Key figures and tables from the main text are reproduced here at enlarged size for readers who find the in-text versions difficult to read. Each item has been rotated 90° counterclockwise to maximize image size on the portrait page. To view, rotate the book 90° clockwise (or tilt your reading device).

Figure 1: The Regulated Sprint Process Overview (detailed walkthrough in Chapter 8)

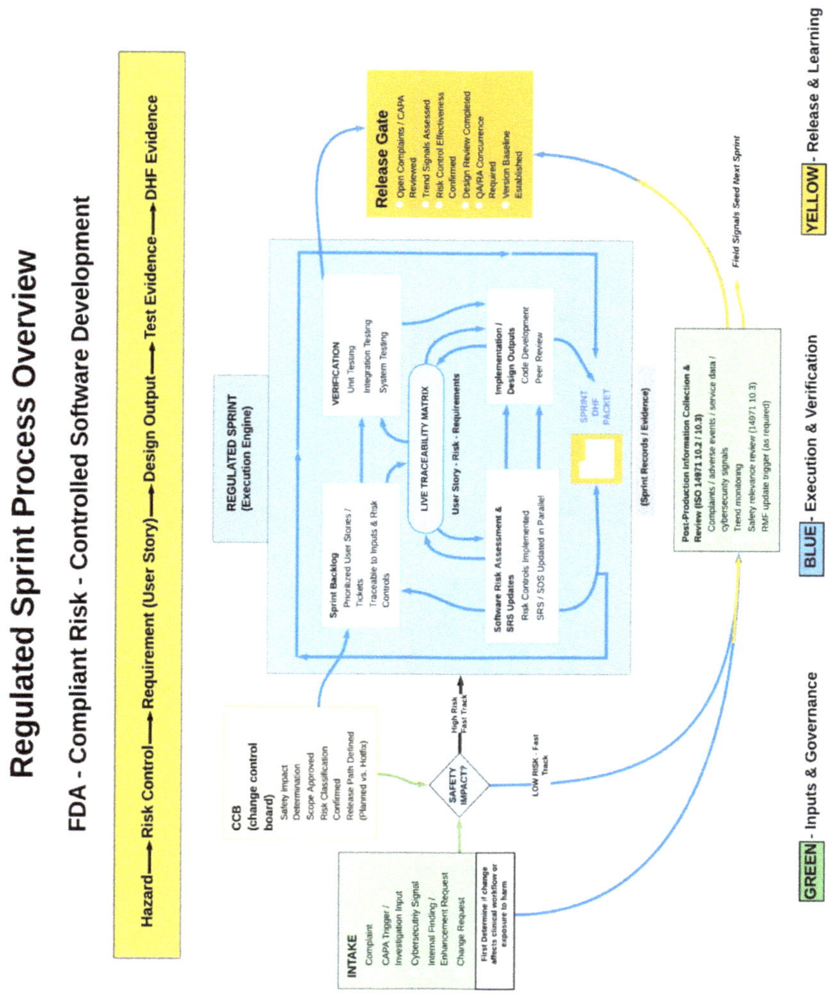

Figure 6-1: ISO 14971 Causal Chain — Device Characteristic to Clinical Outcome with P1/P2 Probability Mapping

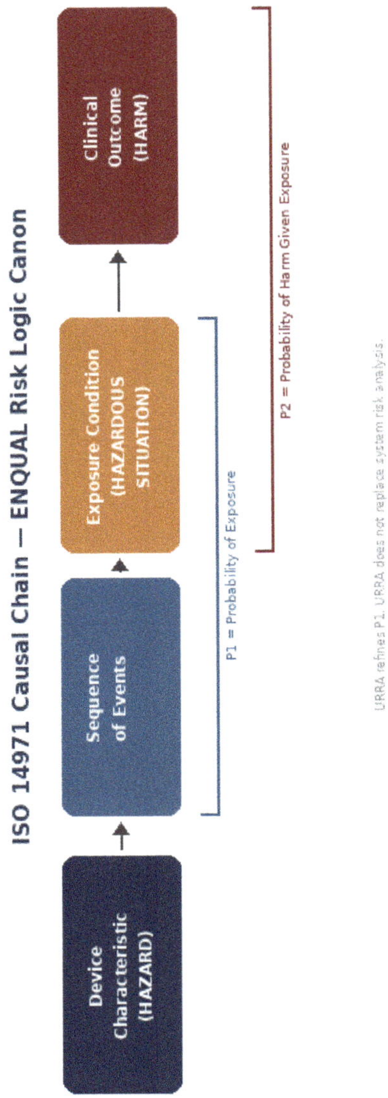

Figure 8-1: Regulated Sprint Board — Sprint 47 showing Safety_Impact classification (a governance triage aligned to risk-based rigor, not the full ISO 14971 risk estimate) for each item

Figure 8-1: Regulated Sprint Board - Sprint 47

Ticket	Title	Safety	CCB Ref	Assignee
BUG-4521	Risk score displays "--" intermittently	MEDIUM	CCB-2026-047	Mike Chen
FEAT-892-2	Alert trigger logic implementation	HIGH	CCB-2026-032	Sarah Lee
FEAT-892-3	Visual alert indicator	HIGH	CCB-2026-032	Sarah Lee
FEAT-892-4	Audio alert tone	HIGH	CCB-2026-032	Tom Park
CHG-104	Dose confirmation dialog	MEDIUM	CCB-2026-041	Mike Chen
CHG-105	Copyright footer update	LOW	CCB-2026-048	Alex Kim

Figure 9-1: CCB Classification Examples — Safety Impact Determination with DQE Rationale

Ticket	Change Description	Classification	DQE Rationale	Key Factor
CHG-101	New drug interaction algorithm	HIGH	Directly affects clinical decision support output	Clinical logic change
CHG-102	Critical alert audio frequency change	HIGH	Affects ability to detect safety-critical notifications	Safety-critical UI
CHG-103	Database query optimization	MEDIUM	No clinical logic change, affects system performance	Reliability impact
CHG-104	Add confirmation dialog before dose	MEDIUM	UI safety enhancement, affects clinical workflow	Workflow change
CHG-105	Update copyright year in footer	LOW	No functional impact, cosmetic only	No safety relation
CHG-106	Fix typo in help documentation	LOW	Non-clinical content, no workflow impact	Documentation only
CHG-107	Change login timeout from 30 to 15 min	MEDIUM	Security change affects clinical workflow continuity	Indirect safety
CHG-108	Add patient photo to verification screen	MEDIUM	Affects patient identification workflow	Identity verification

Figure 9-1a: Why "Form, Fit, Function" Fails for SaMD — The Same Five Changes Evaluated by Each Question

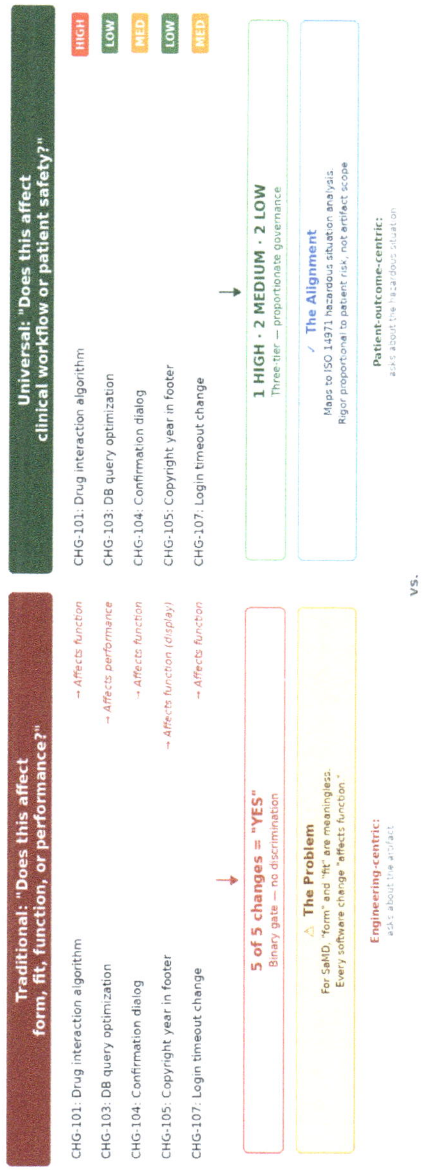

Figure 9-2: Risk-Based Gating Model — Required Controls by Classification Level

Figure 11-1: Bidirectional Traceability — Requirement to Design to Risk to Test

Figure 11-1: Bidirectional Traceability - Requirement to Verification

Req ID	Requirement	Design	Hazard	Test	Result
SRS-4.2	Risk score calculation accuracy	SDS-4.2.1	HA-0052	TC-0089	PASS
SRS-9.1	Critical alert threshold config	SDS-9.1.1	HA-0044	TC-0112	PASS
SRS-9.2	Alert trigger within 500ms	SDS-9.2.1	HA-0045	TC-0113	PASS
SRS-9.3	Visual alert prominence	SDS-9.3.1	HA-0046	TC-0114	PENDING
SRS-9.4	Audio alert distinctiveness	SDS-9.4.1	HA-0047	TC-0115	NOT RUN
SRS-5.1	Dose confirmation required	SDS-5.1.1	HA-0015	TC-0078	PASS

Figure 11-2: Design Control Tracker Dashboard

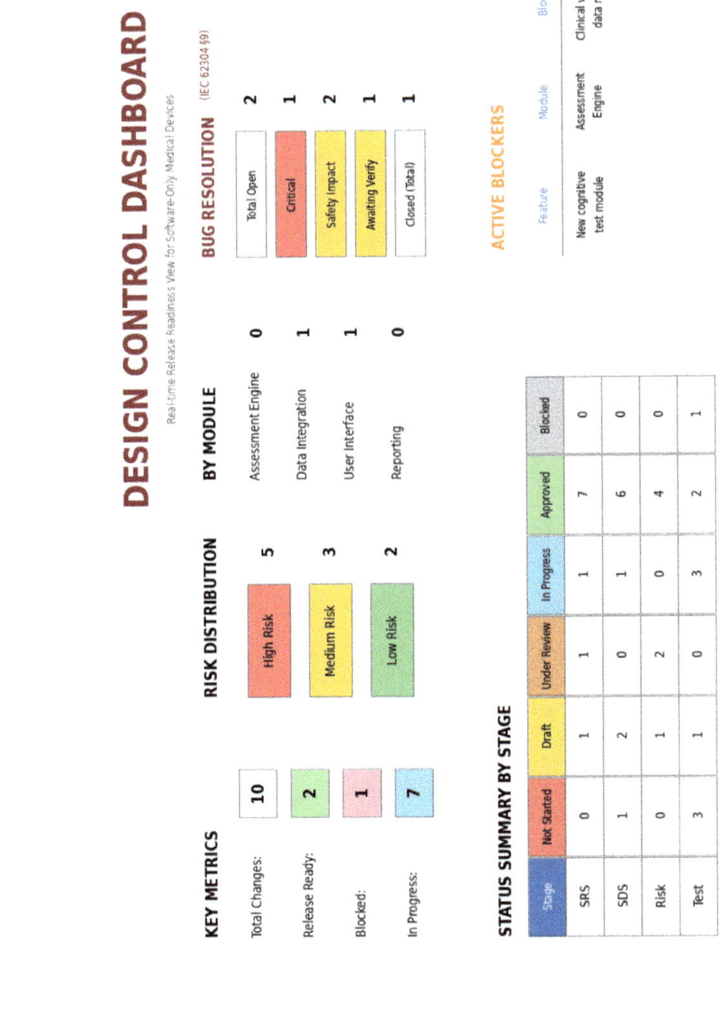

Figure 11-3: Tracker Architecture — Sheet Relationships, Data Flow, and Sprint Selection

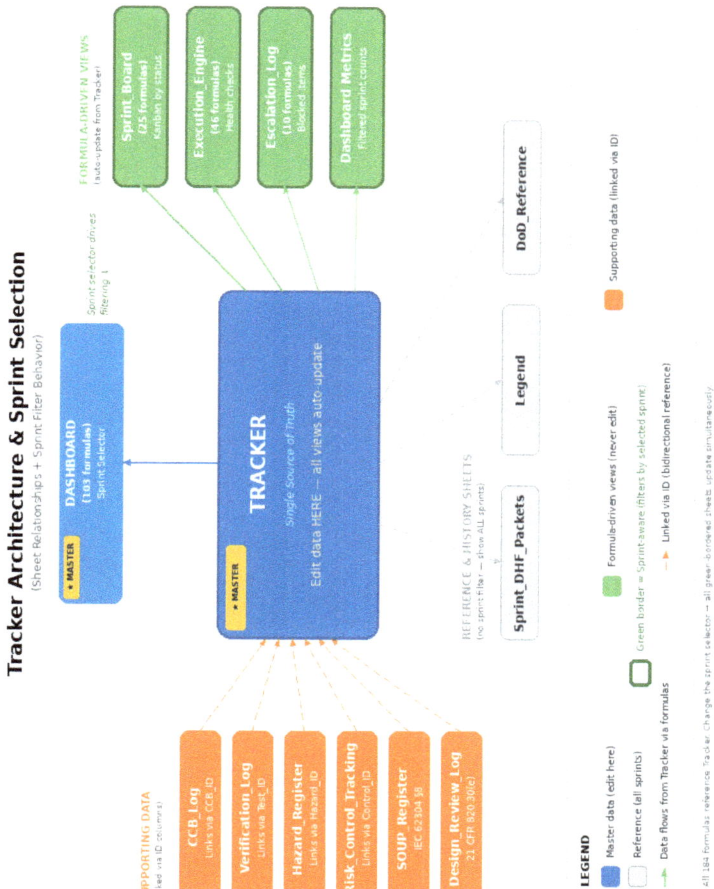

Figure 15-1: Release Gate Dashboard — Sprint 47 showing Evidence Status and Gate Status

Figure B-1: BUG-4521 Complete Lifecycle — From Discovery to Resolution

Figure B-1: BUG-4521 Complete Lifecycle — From Discovery to Resolution

TIMELINE

Time	Actor	Action	Status Change
09:15	QA Tester Sarah	Discovers bug during regression testing	NEW
09:30	Sarah	Moves to CCB Queue per procedure	CCB_QUEUE
10:00	PM Alex	Confirms on CCB agenda for 14:00	CCB_QUEUE
14:00	DQE James	CCB classifies as MEDIUM - UI timing, no clinical data loss	READY
14:30	PM Alex	Assigns to Sprint 47	IN_SPRINT
Day 2 AM	Dev Mike	Pulls into sprint, begins investigation	IN_PROGRESS
Day 2 PM	Dev Mike	Root cause identified: async rendering timeout	IN_PROGRESS
Day 3 AM	Dev Mike	Fix implemented, unit tests added	IN_PROGRESS
Day 3 PM	Dev Mike	Moves to review, links evidence	IN_REVIEW
Day 4 AM	DQE James	Reviews verification evidence, approves	DONE

FINAL TICKET STATE

Ticket ID:	BUG-4521
Title:	Risk score displays "--" intermittently
Type:	Bug
Safety_Impact:	MEDIUM
CCB_Reference:	CCB-2026-047
SRS_Link:	SRS-4.2: Risk Score Display
Risk_Link:	HA-0052: Delayed risk information
Verification_Link:	TC-0089: Risk display verification
Verification_Status:	APPROVED
DQE_Approver:	James Martinez
Resolution:	Fixed - async rendering timeout

KEY OBSERVATIONS

- Same-day CCB classification: Bug discovered 09:15, classified by 14:00
- 4-day resolution cycle: Discovery to Done in under one week
- Complete traceability chain: BUG-4521 → SRS-4.2 → HA-0052 → TC-0089
- Evidence generated during work: Links created as part of normal workflow
- DQE verification at completion: Not during sprint, at verification gate

STATUS FLOW

NEW → CCB_QUEUE → READY → IN_SPRINT → IN_PROGRESS → IN_REVIEW → DONE

Figure H-1: FEAT-892 HIGH Risk Feature — Critical Alert Implementation

EPIC OVERVIEW

Epic ID:	FEAT-892
Title:	Critical Alert Feature
User Story:	As a clinician, I want to be alerted when risk scores exceed threshold
Safety_Impact:	HIGH
CCB_Reference:	CCB-2026-032
CCB_Rationale:	New safety-critical alerting system - affects clinical decision timing
RA_Consultation	Required - potential submission impact reviewed

DECOMPOSED STORIES

Story ID	Title	Safety_Impact	Hazard Link
FEAT-892-1	Threshold configuration UI	MEDIUM	HA-0044
FEAT-892-2	Alert trigger logic	HIGH	HA-0044, HA-0045
FEAT-892-3	Visual alert indicator	HIGH	HA-0046
FEAT-892-4	Audio alert tone	HIGH	HA-0046, HA-0047
FEAT-892-5	Alert acknowledgment workflow	HIGH	HA-0048
FEAT-892-6	Alert history logging	MEDIUM	HA-0049

Figure M-1: Sprint DHF Packet Folder Structure — Self-Contained Evidence Package per Sprint

Figure FF-1: Sprint Status Workflow — Jira States with Phase Color Coding

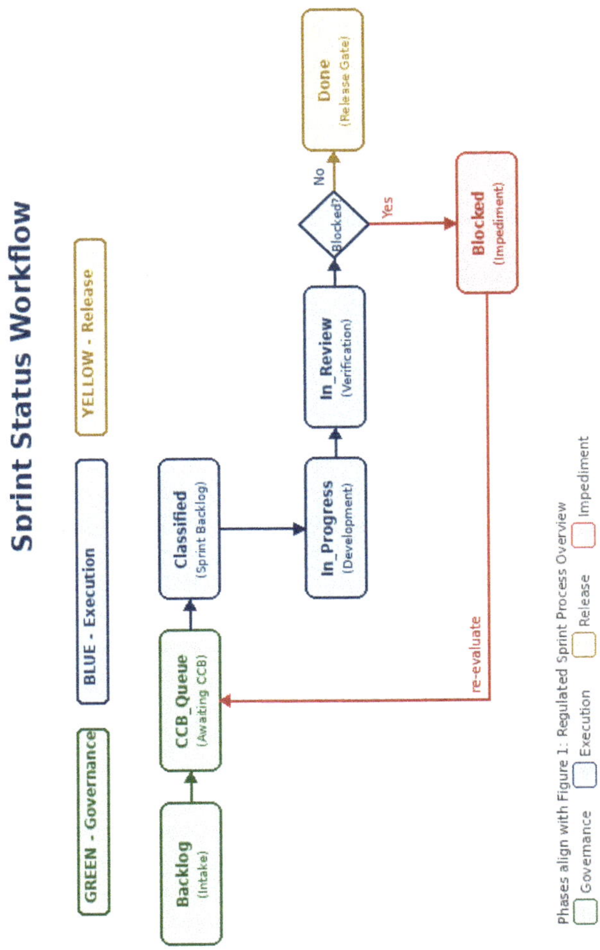

Table Z-1: Compliance Dashboard Metrics (from Appendix Z)

Metric	Current Value	Target	Trend	Status	Action Threshold
CCB Queue Age	(days)	< 3 days	▲ ► ▼	● ● ●	> 5 days
Evidence Completion Rate	(%)	> 95%	▲ ► ▼	● ● ●	< 90%
Documentation Lag	(days)	< 2 days	▲ ► ▼	● ● ●	> 5 days
Verification Cycle Time	(days)	< 3 days	▲ ► ▼	● ● ●	> 5 days
Release Gate Pass Rate	(%)	> 90%	▲ ► ▼	● ● ●	< 80%

Legend: **Trend:** ▲ Improving ► Stable ▼ Declining **Status:** ● On Target ● Warning ● Critical

Enter current values during weekly leadership reviews. Color-code Status based on current value vs. target and action threshold

Book Example Tracker Views (from Appendix HHH) — Large Format

Figure HHH-1: Live Traceability Matrix — Requirement to Test with Risk Level

LIVE TRACEABILITY MATRIX

Real-time view: Requirement → Design → Hazard → Test → Result

Item_ID	Risk	Req_ID	Req_Status	Design_ID	Hazard_ID	Test_ID	Result
STORY-201	MEDIUM	SRS-201	Approved	SDS-3.2.1	HAZ-012	TC-201-01	PASS
STORY-204	LOW	SRS-204	Approved	SDS-4.1.2	—	TC-204-01	PASS
STORY-205	LOW	SRS-205	Approved	SDS-2.3.1	—	TC-205-01	PASS
BUG-301	HIGH	SRS-042	Approved	SDS-5.1.1	HAZ-008	TC-301-01	PASS
STORY-208	HIGH	SRS-208	In Progress	SDS-6.2.1	HAZ-015	TC-208-01	PENDING
STORY-210	HIGH	SRS-210	Approved	SDS-1.4.2	HAZ-003	TC-210-01	IN PROGRESS
BUG-304	MEDIUM	SRS-119	Draft	SDS-7.1.3	HAZ-021	TC-304-01	NOT STARTED

RESULT: PASS IN PROGRESS PENDING NOT STARTED FAIL

Traceability ensures every requirement links to design, hazard analysis, and verified test results

Figure HHH-2: CCB Log — FEAT-892 HIGH Risk Feature Classification

CCB LOG — FEAT-892 HIGH Risk Feature

Classification decisions for Critical Alert epic and child stories

CCB_ID	Date	Item_ID	Title	Risk Level	Severity	Probability	Decision	Verification Req	Note
CCB-2026-032	2026-01-15	FEAT-892	Critical Alert (Epic)	HIGH	Serious	Probable	Approved	Full V&V + Regr	Epic classification drives child story assessment
CCB-2026-033	2026-01-15	FEAT-892-1	Threshold config UI	MEDIUM	Marginal	Probable	Approved	Unit + Integ	
CCB-2026-034	2026-01-15	FEAT-892-2	Alert trigger logic	HIGH	Serious	Probable	Approved	Full V&V	
CCB-2026-035	2026-01-15	FEAT-892-3	Visual alert indicator	HIGH	Serious	Probable	Approved	Full V&V	
CCB-2026-036	2026-01-15	FEAT-892-4	Audio alert tone	HIGH	Serious	Frequent	Approved	Full V&V + Regr	
CCB-2026-037	2026-01-15	FEAT-892-5	Alert ack workflow	HIGH	Serious	Occasional	Approved	Full V&V	
CCB-2026-038	2026-01-16	FEAT-892-6	Alert history logging	MEDIUM	Marginal	Remote	Approved	Unit Test	

Epic-level classification cascades to child stories | Each story assessed independently at CCB

Figure HHH-3: CCB Classification Examples with P1 × P2 Risk Matrix

Figure HHH-4: Sprint DHF Packets — Release-Ready Evidence Package

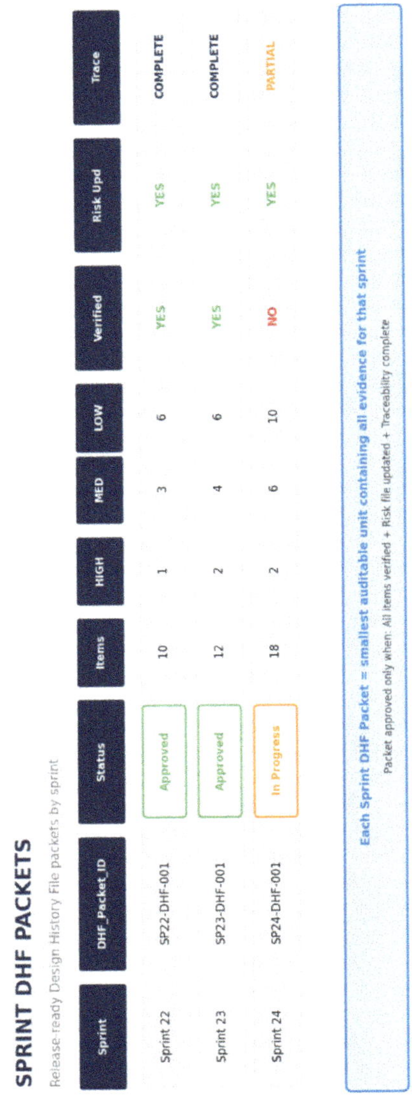

SPRINT DHF PACKETS

Release-ready Design History File packets by sprint

Sprint	DHF_Packet_ID	Status	Items	HIGH	MED	LOW	Verified	Risk Upd	Trace
Sprint 22	SP22-DHF-001	Approved	10	1	3	6	YES	YES	COMPLETE
Sprint 23	SP23-DHF-001	Approved	12	2	4	6	YES	YES	COMPLETE
Sprint 24	SP24-DHF-001	In Progress	18	2	6	10	NO	YES	PARTIAL

Each Sprint DHF Packet = smallest auditable unit containing all evidence for that sprint

Packet approved only when: All items verified + Risk file updated + Traceability complete

Figure HHH-5: Live Traceability Matrix — Bidirectional Links with Status

LIVE TRACEABILITY MATRIX — Bidirectional Links

Requirement → Design → Hazard → Test → Result

Item_ID	Risk	Req_ID	Req_Status	Design_ID	Hazard_ID	Test_ID	Result
FEAT-892-1	MEDIUM	SRS-9.0	Approved	SDS-9.0.1	HA-0044	TC-0111	PASS
FEAT-892-2	HIGH	SRS-9.1	Approved	SDS-9.1.1	HA-0044	TC-0112	PASS
FEAT-892-3	HIGH	SRS-9.2	Approved	SDS-9.2.1	HA-0045	TC-0113	PASS
FEAT-892-4	HIGH	SRS-9.3	Approved	SDS-9.3.1	HA-0046	TC-0114	PENDING
FEAT-892-5	HIGH	SRS-9.4	Approved	SDS-9.4.1	HA-0047	TC-0115	IN PROGRESS
FEAT-892-6	MEDIUM	SRS-9.5	Approved	SDS-9.5.1	HA-0049	TC-0116	PASS
BUG-4521	MEDIUM	SRS-4.2	Approved	SDS-4.2.1	HA-0052	TC-0089	PASS
CHG-104	MEDIUM	SRS-5.1	Approved	SDS-5.1.1	HA-0015	TC-0078	PASS

RESULT: PASS PENDING IN PROGRESS NOT STARTED FAIL

Figure HHH-6: Release Gate Dashboard — Sprint Readiness with Gate Status

www.ingramcontent.com/pod-product-compliance
Ingram Content Group UK Ltd.
Pitfield, Milton Keynes, MK11 3LW, UK
UKHW020721041026
15590UKWH00002B/792

9 798994 787700